Explorations in the Life of Fishes

Harvard Books in Biology, Number 7

Editorial Committee: Ernst Mayr, Keith R. Porter,
John R. Raper, Edward O. Wilson

Explorations in the Life of Fishes *N. B. Marshall*

Harvard University Press, Cambridge, Massachusetts 1971

London: Oxford University Press

Preface

This book began with lectures given at Harvard University during the spring term, 1963. Parts of these lectures were used to prepare the H. R. MacMillan Lectures in Fisheries, which I gave at the University of British Columbia in January and February 1964. Once more, I would like to express warm thanks to my sponsors, the Oceanographic Committee and the Museum of Comparative Zoology, Harvard University, the Woods Hole Oceanographic Institution, and the Institute of Fisheries, U.B.C.

Since 1965 I have expanded the content and scope of my interests, particularly through further research. In late 1967 and early 1968 I used some of this work to give a series of lectures at the University of Miami and the Institute of Marine Science where I was a visiting professor. Again I wish to reiterate my gratitude.

The title of this book expresses my endeavors and I believe the contents will be of general interest to biologists. Deep-sea fishes do not live in a vacuum. Moreover, one cannot discuss convergent evolution in fishes without at least considering the fish-like features of other aquatic organisms. To help those not familiar with fishes, there are numerous illustrations and an appendix outlining the classification and evolution of teleosts. (For reasons of conformity, I have followed the classification used in the Pisces section of the Zoological Record, published each year by the Zoological Society of London).

Finally, thanks are due to diverse helpers. Many have provided lively comment and discussion on my lectures and manuscript. My wife has once more contributed greatly, at the same time keeping a critical eye on the two illustrators.

Contents

Illustrations

by N. B. Marshall and Lesley Marshall

Explorations in the Life of Fishes

1 The Success of Teleostean Fishes

During middle Mesozoic times one or more kinds of holostean fishes evolved into teleosts. By the end of the Cretaceous period, about 100 million years later, teleosts had become the dominant fishes of the hydrosphere. Today, after a further 70 million years of evolutionary radiation there are some 20,000 species. The next most diverse fishes are the sharks (about 225 species) and rays (about 350 species). Except perhaps for the lampreys (30 species)[1] and hagfishes (15 species), the other modern fishes are relics of once diverse groups. There are six species of lungfishes, one coelacanth, 36 palaeoniscoids (bichirs 11, sturgeons 23, and paddlefish 2), 8 holosteans (1 bowfin, 7 garpikes), and 25 chimaeroid fishes (Holocephali). (See Figure 1.)

Numbered in individuals as well, teleosts are overwhelmingly the predominant fishes. They also exploit many more kinds of living spaces than do other fishes. In fresh waters they range from torrents to still or sluggish waters, and from puddles to the deeper parts of Lake Baikal (about 1,600 meters). Temperatures around them extend from zero to 100°F, or so, of hot springs; nor have stagnant swamps and pitch dark underground waters been barriers to their radiations. In the ocean they live from surf-swept shores to slow flowing waters at depths of 9,000 meters or more, which means that they are also adapted to wide ranges of light, temperature, salinity, oxygen concentration, and so forth. In brief, teleosts have succeeded in almost every kind of living space in the hydrosphere. They are, so to speak, the birds of aquatic space. Indeed, A. J. Marshall (1960) has drawn attention to certain striking parallels between birds and teleosts, particularly in their reproductive behavior.

Teleosts, then, have succeeded very well in these respects: they are represented by vast number of individ-

1. A lamprey (*Mayomyzon*), very like modern forms, has been discovered recently in middle Pennsylvanian deposits of northeastern Illinois. Thus the evolution of lampreys may well go back to Palaeozoic times (Bardack and Zangerl, 1968).

The Life of Fishes

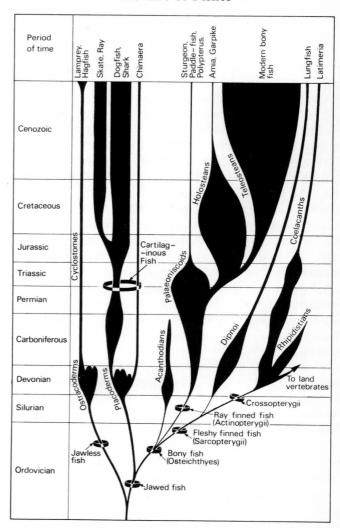

1 Evolutionary relationships of fishes. (From A. S. Romer, *The Procession of Life*, 1968, courtesy of Weidenfeld and Nicolson.)

uals belonging to very many species that cover a great range of living spaces. Cephalopods, the most fish-like of the invertebrates, have not, as Wells (1962) reminds us, colonized fresh waters, partly because their kidneys are unsuited to the rapid filtration that is needed in a salt-poor environment. Moreover, compared to the hemoglobins, the hemocyanin system, which the cephalopods retained, has limitations as a carrier of oxygen. Most cephalopods need well-aerated waters.[2] "Specializing, as it were, in one particular mode of life, the group has produced some of the most powerful predatory animals in the sea. And it looks as though in doing so they have become specialized to the point where their own structure and physiology will preclude any further adaptive radiation" (Wells, 1962). Apart from the type of blood pigment, much of this assessment applies to the elasmobranchs, though a few species have spread into fresh water. Indeed, cephalopods are more versatile than elasmobranchs in their ability to move forwards or backwards, and some of them, like a few species of sharks, have the means for attaining neutral buoyancy (Denton, 1963).

Birds and mammals were also evolving in Mesozoic times, and, like the teleosts, became dominant groups by the end of the Cretaceous period. Now, the diagnostic features of the Aves and Mammalia, give considerable insight into the reasons for their success, but the same cannot be said for the Teleostii. Definitions of these fishes stress their loss of certain holostean characters, such as ganoin, rostral bones, lower jawbones, fulcral scales, gular plates, and so forth. As positive features, there is a greater mobility in the upper jawbones, the vertebral centra are typically well ossified, and the caudal fin, unlike that of most holosteans, is compact and homocercal in form.[3] Concerning their soft parts, some of the most impressive advances are in the design of the central nervous system, the swimbladder, and the reproductive

2. Certain species, for instance, *Vampyroteuthis infernalis*, live in oxygen-minimum layers.

3. Patterson (1967) provisionally defines the teleosts on the typical structure of the caudal fin: they are "actinopterygian fishes in which the lower lobe of the caudal fin is primitively supported by two hypurals borne on a single centrum, and in which the primary squamation of the caudal axis is modified into elongate urodermals, the anterior being in contact with vertebral centra and overlain by scales."

system. But it is not possible to express such trends in a group definition. The most striking diagnostic feature of the teleosts is in their phenomenal success. (See Appendix.)

SIZE

Thinking especially of the ubiquity of insects, one begins to appreciate the biological advantages of smallness. The sizes of teleosts are also in their favor. They range from half an inch to about 15 feet in length, but the great majority fall between lengths of one inch and one foot. An average-sized teleost is probably close to a span of 6 inches.

The biological significance of size in any particular species is obscure, though a comparative study of the sizes of related sympatric species can be illuminating (Hutchinson, 1958). It is easier to gain some appreciation of the importance of size at the upper and lower limits of the size range in a group of organisms. The smallest known teleost is *Pandaka pygmaea*, a goby that grows to a length of 11 mm. Another Philippine goby, *Mystichthys luzonensis*, which is almost as small as *Pandaka*, has been studied carefully by Te Winkel (1935). The cells of this species are of much the same size or a little smaller, compared to those of larger teleosts. Even so, very small animals have certain advantages in surface to volume ratios. Organs with functions depending on surfaces, for example, gills, kidneys, and gut, can be relatively small and simple, as indeed they are in *Mystichthys*. Such organs must be comparatively easy to run. The main biological problem concerns volume-dependent parts of the body. In particular, the eggs of *Mystichthys* are comparaparatively large (diameter 0.37 to 0.4 mm), there being 20 to 40 per female. Two larger gobies, *Acentrogobius neilli* and *Gobionellus boleosoma*, produce eggs of much the same size. These would seem to be the smallest eggs, presumably to produce viable larvae, that have been evolved among teleosts. This lower limit evidently sets a limit to the smallness of the females. The fecundity of a fish 5 mm in length would be very reduced. As things are, both *Pandaka and Mystichthys* are the basis of important fisheries.

Smallness in fishes may well be limited by another factor, swimming speed, which decreases with body length.

The Success of Teleostean Fishes

Looking at Zenkevitch's (1945) graphs of body size in relation to velocity for certain organisms, Shuleikin noticed that below a length of about 1 cm "fishes began to yield to Copepoda the maximal velocity of motion. This is why in the course of evolution the motor of fishes has come to the foreground only at a definite stage of length increase." Fishes are well organized for life and motion in water, but they must evidently be 1 cm or more in length to take advantage of this organization.

Lastly, eyes, ears, and brain cannot be diminished beyond certain sizes if they are to function properly (Haldane, 1928). This may well be seen in *Mystichthys*, in which the above organs are relatively large.

At the other end of the size spectrum, the most stringent limiting factor is the stress of finding enough food to maintain great bulk requirements. The largest whales obtain the greatest possible biomass of food by consuming small herbivores, such as krill and *Calanus*. The largest fishes, the whale shark, the giant manta ray, and the basking shark, depend largely on planktonic animals and small fishes. Moreover, the first two species are confined to warm water, where plankton production is more or less continuous throughout the year. The basking shark, which lives in temperate seas, loses its gill rakers during the lean colder months, when it is presumed to rest until the next plankton season, by which time it will have acquired a new set of rakers.

Hutchinson and MacArthur (1959) have argued on theoretical grounds that at a particular level in the food web and above a certain lower limit of size, there are likely to be more species of small and medium-sized animals than of larger ones. Evolutionary opportunism for animals of modest size has been greatest where there is much trophic and ecological diversity. One has only to think of the many species of small teleosts that live in coral reefs and atolls. Some of the smallest, particularly gobies and eleotrids, live among the branches of coral, while others, such as the smaller pomacentrids, use coralline bolt-holes to escape from their enemies. Larger species, notably holocentrids and moray eels, stay in coral shelters by day and use the cover of night to feed. There are also diverse niches along rocky, weed-covered shores, and again, many of these are occupied by small gobies, blennies, clingfishes, and so forth. The very smallest sharks and rays are too large for such niches. *Squaliolus*

(6 to 12 inches) and *Euprotomicrus* (12 inches), the smallest sharks, are larger than a middling-sized teleost, while the smallest rays have a disc width of about 2 inches. A mean-sized shark is probably near 6 feet in length, while the same figure for rays is close to 3 feet. Even where niche diversity is limited, as in the mid-waters of the ocean, there are many more species of teleosts than of elasmobranchs. Considering only a predatory group, the one hundred-odd species of ceratioid anglerfishes contrast strongly with the few species of mid-water sharks. But some of the ceratioids at least, like certain other deep-sea fishes, break the rule that the predator is larger than its prey. By breaking the rule, and most ceratioids are rather small, they seem to realize the Hutchinson-MacArthur expectation.

There is a further and more obvious connection between the sizes of fishes and their food. In the hydrosphere, particularly in the ocean, there is an abundance of small food organisms, suitable for animals of modest sizes and needs. No doubt the protochordate ancestors of fishes were ciliary-mucous feeders, and such microphagous habits, as Morton and Miller (1968) stress, are extremely efficient, considering the very small expenditure of energy involved. By evolving jaws, fusiform bodies, and extensive axial muscles, which are costly to maintain, fishes were able to exploit larger size-ranges of food organisms. This is especially true of the elasmobranchs: indeed, the only small forms with screens for retaining planktonic food are the smaller manta rays. But many teleosts evolved means of imitating the protochordates, so staying near the more productive, sunny end of the food pyramid. Epibranchial organs, designed for microphagy, are present in many of the "lower" teleosts, particularly the clupeids, which in terms of biomass are outstandingly successful. Even so, teleosts with a more advanced branchial complex are able, by one means or another, to make use of the wide availability of small food organisms. One has only to think of the gobies.

There is a final advantage of modest size. The smaller the organism, the shorter the time it needs to reach reproductive maturity (generation time; Bonner, 1965). Very successful land-dwelling animals, insects, passerine birds, and rodents, tend to have relatively short lifespans, and Wynne-Edwards (1962) argues that the advantages of short cycles have contributed to the success of these groups. If this is so, then the same arguments can be used in water for the teleosts. The quicker the succession of generations, the more are the chances of viable genetic changes. But there is no general correlation between generation time and the rate of evolution (Simpson, 1953). Hutchinson (1965) puts it thus: "Two extreme ways of evolution in relation to time are possible. Since natural selection will proceed faster when generations succeed each other faster, one way is the evolution of progressively smaller and more rapidly reproducing organisms. However, the smaller an organism, the less it can do. An alternative path gives larger, slower-reproducing organisms in which, when a nervous system capable of learning is developed, a premium is put on experience." It would be fair to say that most teleosts have struck a compromise between these extremes.

BUOYANCY

Most teleosts that swim and hover with ease in aquatic space have a swimbladder, which tends to keep their weight at the vanishing point.[4] This hydrostatic kind of swimbladder is characteristic of teleosts, and in view of its primary function, it is not surprising to find this organ is lost or reduced in species that are closely bound to the bottom.

A just-buoyant swimbladder must have played a leading part in the evolutionary success of the teleosts. Besides its wide specialization as a hearing aid and a sound producer, it enables a fish to save energy. To keep at one level, a freshwater teleost deprived of its swimbladder must "tread water" with a force equivalent to some 7 or 8 percent of its weight in air. For a marine fish this figure is about 5 percent. E. J. Denton and T. I. Shaw have calculated that "for some active pelagic fishes a force of 7% of the body weight would enable a fish to *sustain continuously* a horizontal velocity of 43% or 53% (depending on whether the flow is laminar or turbulent) of that velocity which it would have if it exerted *continuously* a force of 25% of its body weight, a force fishes *seldom* exert"

4. The swimbladder is regressed or absent in numerous mesopelagic and bathypelagic species, but such species, notably by reduced skeletal and muscular system, are close to being neutrally buoyant (Denton and Marshall, 1958).

(Denton and Marshall, 1958). Sharks and rays, which have no swimbladder, must exert a downward force equivalent to about 5 percent of their weight if they are to swim at one level. Rays that spend much of their life in aquatic space move their wings, which have a hydrofoil shape in section, in a bird-like way, while the pectoral–caudal fin coupling of sharks is designed to produce a strong lift force when the fish is in motion.

To achieve perfect buoyancy, the teleosts acquired a hydrostatic swimbladder and also evolved a lighter kind of skeleton. Not only did they shed heavy ballast in the form of ganoin, but their bones have a more open, lighter structure than those of holosteans. Holostean bone, as Rayner (1948) stresses, is heavy and cancellous in texture, as is the skeleton of the early palaeoniscoids and tassel-finned fishes. In fact, the early crossopterygians, dipnoans, and actinopterygians, had low-set pectorals and a markedly heterocercal tail. In all probability, these fishes were heavier than water, in spite of their lungs or swimbladder. Some idea of the problems of buoyancy in a heavily armored fish may be got from a modern genus, *Polypterus*, which is closest to the ancient palaeoniscoids. The lungs of *P. palmas* and *P. delhezi* are 9.8 and 12.4 percent of their respective body volumes, but these figures would need to be 12.4 and 13.7 percent to give neutral buoyancy. In presenting these data, Alexander (1966) states that the heavy ganoid scales of *Polypterus* are 10 percent of the body weight in each species. The scales of a teleost (*Rutilus*) are much lighter, amounting to 4 percent of the body weight. Alexander proceeds to argue that the early Osteichthyes living in fresh water would have required a very large swimbladder to achieve neutral buoyancy. This requirement can be met, as *Lepisosteus*, a heavily armored holostean, shows today.

Yet a large swimbladder has disadvantages; for the greater its capacity, the more effort a fish must exert to keep station after a given rise or fall above a level of equilibrium. As Alexander (1966) concludes, a fish with a small swimbladder has advantages over one with a larger organ. Teleosts possess a relatively small swimbladder. In fact, the poise of a physoclistous teleost is hardly disturbed after a quick climb—away from a depth of neutral buoyancy—that involves as much as a 25 percent increase in the volume of the swimbladder. Thus, the smaller the swimbladder, the more readily can a fish regain neutral buoyancy—if it has the means to do so—after a dive or an ascent. Teleosts have evolved such means (Jones and Marshall, 1953). Indeed, once they had acquired a small-capacity swimbladder, the way was open for the evolution of buoyancy-readjusting systems.

The evolution of a small swimbladder must have been correlated closely with that of a light skeleton. We have seen just how much weight can be lost by the scales. The membrane bones of the skull also lost their ganoin and, like every element of the inner skeleton, acquired an open, strutted structure.[5] More precisely, the apatite complex, which has a high density and gives bone a strength-to-density ratio that may be more than half that of steel, tends to be concentrated along struts. Between the struts the bone is relatively thin. Plate-like bones, such as those of the skull roof and the gill covers, are radially strutted. Vertebral centra may be braced by fore-and-aft struts, and so forth. Bone has been pared away but the skeleton is still strong—strong enough to work with axial muscles that are from 40 to 70 percent of the body weight in active kinds of teleosts.

As man with an aqualung has found, being just buoyant not only enables one to stay at a particular level with virtually no effort but gives great freedom for underwater maneuver. Concerning the second aspect, teleosts were well endowed by inheriting flexible kinds of fins from their holostean ancestors. In the dorsal and anal fins, each ray, which moves on its own basal element, is controlled by elevator, depressor, and inclinator muscles. Thus the entire fin can be raised, lowered, and undulated. Such a fin can be quickly hoisted as a pivot when making a turn or folded down to reduce resistance when traveling at speed. Undulations passing down a median fin provide thrust, which can also be got if the fin is flapped from side to side.

It is reasonable to associate the homocercal form of caudal fin with a just-buoyant body (Harris, 1953). The heterocercal caudal fin of sharks, coupled with the down-warped pectoral fins, provide lift for a body that is heavier than water. This coupling was found in the early bony fishes, which are likely to have been negatively buoyant, in spite of their swimbladder. A perfectly

5. The strutted patterns of these bones remind one of certain of the reinforced concrete structures designed by Nervi (1965).

buoyant fish obviously needs no lift, and in the teleosts the caudal became symmetrical in profile and the pectorals were freed from their role as hydroplanes. They were free to evolve in other directions. In the more advanced teleosts, particularly the percomorphs, the pectorals are set laterally on the shoulders and can be used as brakes and oars. To maintain stability during braking, the pelvic fins have moved forward close under the pectorals with which they act in concert. In brief, teleosts evolved very versatile fin systems, which were closely dependent on the attainment of neutral buoyancy and its homeostatic control. To appreciate this today, one has only to think of the diversity of teleosts in coral reefs and atolls. Many species use fins other than the caudal member for locomotion. When hovering, feeding, courting, breeding, and so forth, all manner of precise fin movements succeed one another. Reproductive behavior in particular depends on signals flashed by the fins. No such fin versatility is to be found among elasmobranchs and the modern relics of the earlier Actinopterygii.[6]

JAWS

The early teleosts acquired a ball-and-socket joint between the palatine and maxillary bones. In more advanced forms, these mobile maxillae, though an essential part of the jaw mechanism, are excluded from the gape by the premaxillae. In still more advanced teleosts, the premaxillae became protrusible, which "enables the fish to project its upper jaw toward food with great rapidity while extending the orobranchial chamber further forward than was possible by maxillary movement alone" (Schaeffer and Rosen, 1961). Protrusible jaws are found in well over half of the teleosts, notably in cyprinids and percomorphs. Such jaws have been adapted to a wide range of feeding habits, though in some pelagic predatory types, such as the scombroids and trichiuroids, the jaws are fixed. Protractile jaws are particularly fitted for taking food from the bottom or from confined spaces.

Most of the elasmobranchs are predatory fishes. This

6. *Amia*, which comes closest to teleosts in fin versatility, also has light, ganoin-free scales and bones.

position in the food web is linked with their limited adaptive radiation. The very simplicity and versatility of their digestive system, together with means for storing food reserves in a large liver, "permits a wide choice of food and allows the shark to survive long fasts between periods of gluttony" (Springer, 1960).

REPRODUCTIVE ASPECTS

The production of few and relatively large young, which are soon ready to fend for themselves, has also helped the cartilaginous fishes to survive. Most teleosts start life as small larvae, and many eggs are produced to cover the great mortality during their early existence. In the sea the great innovation was the evolution of buoyant eggs, which are produced by diverse shelf-dwelling and deep-sea teleosts. During the latter stages of maturation, follicle cells secrete dilute body fluids, containing half or less of the salt content of sea water, into the developing eggs. Part of this store of buoyant fluid is retained beneath the skin of the larvae, and they thus float with ease. (These buoyancy chambers are soon replaced by the swimbladder.) Denton (1963) even suggests that the primary biological significance of dilute body fluids in marine teleosts resides in their use for producing buoyant eggs.

Despite the high mortality, the broadcasting of many buoyant eggs and larvae by currents enables a species to cover and exploit its living space, and even, when possible, to extend it. As Mayr (1965) remarks: "That numerous species of marine organisms practise this type of reproduction, many of them enormously abundant and many of them with an evolutionary history going back several hundred million years, indicates that this shotgun method of thrusting offspring into the world is surprisingly successful." The essence of survival is, of course, the maintenance of local populations, and here the teleosts display considerable adaptability. In the northwestern Pacific, for instance, where strong currents are liable to carry young to unfavorable areas, there is an increased fecundity of certain species (for example, of capelin, cod, and herring) compared to their nearest relatives in the North Atlantic (Zenkevitch, 1963).

SENSE ORGANS AND CENTRAL NERVOUS SYSTEM

The great adaptability of teleosts is reflected by their sense organs and central nervous system. The keen olfactory sense of sharks and rays is well matched by that of eels, but in teleosts as a whole there is every gradation of microsmatic to macrosmatic kinds of olfactory system. Microsmatic species usually have well-developed eyes. There is wide variation, too, from diurnal to noctural kinds of eyes. Though some sharks and rays, such as certain species of carcharinids, *Lamna*, *Mustelus*, *Squatina*, and *Myliobatis*, do have cones in their retinae, the great emphasis in cartilaginous fishes is toward rod-rich, nocturnal eyes. Gilbert (1963) concludes: "The rod-rich, cone-poor retina of sharks with its high ratio of visual cells to bipolar and ganglion cells, comprises an eye that has low visual acuity but high sensitivity and that can readily detect an object against a contrasting background in even the dimmest light. The importance of such an eye to a fish that usually feeds at night is obvious." Moreover such eyes have limited the adaptive radiation of cartilaginous fishes. For instance, these fishes do not have central and foveal areas in the retina, such as are found in diverse teleosts. To use these special visual centers calls for precise control of buoyancy and movement, which brings us back to earlier discussion of teleost capacities.

"In most sharks and rays," concludes Aronson (1963), "olfaction appears to be the dominant sensory system. The lateral line is the one other system that develops to any degree in sharks. From an anatomical point of view, the central connections between the olfactory and lateral line systems are remote . . . Also, the connections between the optic lobes and the olfactory centres and between the optic lobes and the static system are much more limited than in teleosts."

Concerning the forebrain, recent study of actinopterygians by Nieuwenhuys (1962) is revealing. The basal region (area ventralis) of the forebrain has much the same organization in bony fishes, but there is much more variation in the structure of the pallial region (area dorsalis). In fact, there is progressive pallial differentiation in the series palaeoniscoids (*Polypterus*)–ganoids (Chondrostei and Holostei)–teleosts, even leading in some teleosts to the formation of a pallial cortex. In general the teleosts show the emergence of a nonolfactory center in the pal-

The Success of Teleostean Fishes

lium (connecting with the ventral thalamus and hypothalamus) and the elaboration of links between the pallium and more caudal parts of the brain. Nieuwenhuys correlates these developments with experiments on teleosts that show the forebrain to be involved in "initiative," schooling, breeding behavior, and so forth.

In fact, the teleost brain is more elaborate and plastic than is the brain of cartilaginous fishes. Olfactory, visual, and autonomic processes are much more closely related through neural pathways to lower motor centers in the spinal cord of teleosts. The Mauthnerian system, which is found in most bony fishes and *Chimaera*, is lacking in sharks and rays. The two giant Mauthner neurons are linked to parts of the acoustico-lateralis, optic, and cerebellar systems, and to somatic motor neurons by way of decussating giant fibres that run the length of the spinal cord. By this means the swimming muscles are brought massively into play in a rapid escape movement. Such quick and powerful locomotion is needed in a dense medium. Though the pectoral fins of rays are not built for sudden and rapid acceleration, it is curious that sharks are without a Mauthnerian system. Aronson (1963) suggests that this is why sharks are incapable of rapid acceleration.

Though the cerebellum may be large and convoluted in some of the larger sharks, the complexity of this part in cartilaginous fishes does not attain the teleost level. "Cerebellar cortex is the dominant regulator of the action of all skeletal muscles and so was called by Sherrington the "head ganglion" of the proprioceptive system. The pattern of action of these muscles is determined elsewhere and the cerebellum has nothing to do with that. As the engineer would express it, it is a servo-mechanism. It controls the execution of these acts by regulation of the strength and timing of contraction of the separate muscles involved. This regulation is entirely involuntary, for it is effected on the outgoing or efferent side of the nervous circuits. The size of the cerebellum varies with the amount and nature of muscular activity. It is small in sluggish species and very large in active fishes, in birds, and in all other vertebrates with rapid movements requiring accurate control" (Herrick, 1961, p. 369). The cerebellum gathers its data from muscle spindles, the acoustico-lateralis, visual, and other exteroceptive systems.

In teleosts, Bianki (1963) suggests that the cerebellum coordinates its motor role with swimbladder functions, such as those involving sensitivity to pressure. If this is so, and considering that teleosts have very versatile fin systems, we might expect them to have a somewhat more elaborate cerebellum, than the elasmobranchs. But the latter fishes even lack the valvula part of the cerebellum, which in teleosts is concerned with visual, static, gustatory, and other sensory information. There are also more intricate connections between the midbrain, diencephalon, and cerebellum in teleosts.

To conclude, one readily agrees with Aronson (1963) that "the teleostean brain is highly variable. Pronounced structural differences are seen even among species of a single genus. The elasmobranch brain, on the other hand, is relatively uniform. Although some variability is seen in the forebrain and cerebellum, this never reaches the proportions found among teleosts. This variability in brain structure is correlated with diverse habits in many families of bony fish. Included here are variations in locomotion, orientation, social and sexual behaviour and, above all, in feeding patterns. Neural adaptability has made it possible for teleosts to enter almost all of the available ecological niches, whereas elasmobranchs with their uniform, less adaptable and more diffuse nervous systems, are confined for the most part to the relatively uniform conditions of more or less open waters."

Teleosts thus lead richer lives than other fishes. One aspect of this has already been mentioned: the flashing of signals by mobile and versatile fins. Bodily postures, the flaring of gill covers and membranes, and changes in color pattern may also be involved in signaling. Sonic signals, whether stridulatory or coming from the swimbladder, are also integral to the lives of many teleosts. Sounds, like visual signals, are associated with attack, defense, social, and sexual activities. Sharks and rays, with a cartilaginous skeleton and no kind of swimbladder, are without mechanisms of sound production (Marshall, 1962). But the diverse forms of "language" in teleosts need hardly surprise us, for they are some mechanisms to preserve the genetic identity of diverse species.

A good many teleosts are involved in symbiotic, commensal, and inquiline associations with other animals. Even more striking, about a quarter of teleost species school during some part of their lives. Schooling, whether to reduce predation or keep together reproductive potential, is clearly of great significance in the lives of teleosts. Few elasmobranchs form schools.

THE TWO MAIN GROUPS OF TELEOSTS

Eventually, though, any discussion of the success of teleosts must be particularly concerned with the Ostariophysi and Percomorphi. These orders contain over half the species of their subclass. The Ostariophysi, which have been outstandingly successful (about 4 species out of every 5) in freshwater habitats, have two special features. By virtue of the chains of ossicles linking the ears to the swimbladder they are, to cite Dijkgraaf (1960), hearing specialists. They have also evolved a remarkable form of olfactory signaling.

When compared to species without a swimbladder hearing aid, members of the Ostariophysi display sharper and more discriminating powers of hearing. They may also hear over a wider range of pitch. Close appreciation of the significance of good hearing to any ostariophysan has yet to be attained, though many catfishes have evolved sound-making devices, which are also present in certain characids and cyprinids (Marshall, 1962). There is much to be discovered, and most likely we shall come to realize the vital part played by sound signals in the group, many members of which live in murky tropical waters.

Means of olfactory signaling reside in special club cells in the epidermis, which produce pheromones. These chemicals are released into the water when the skin of a fish is injured. On perceiving the pheromone of an injured individual, members of the same or closely related species rapidly escape. Active club cells are present in the skin of newly hatched fishes, but the fright reaction is not developed until a later stage. Injury to a young fish will thus scare related adults but not the contemporaries of the wounded one. Cannibalism is much less likely. Moreover, "Like other groups of fish, the Ostariophysi can be conditioned rapidly to odors, and the innate fright reaction (*Schreckreaktion*) can be elaborated and intensified through conditioning. For example, when a pike attacks and alarms a school of minnows, the odor of the pike also becomes associated with the fright reaction; pike odor then takes on a meaning which it did not have before the encounter. Reactions to pike odor are not innate although the *Schreckreaktion* is. Likewise, visual stimuli may become associated with the fright reaction to give it added stimulus in certain situations. This fright reaction is confined to the Ostariophysi and has considerable biological importance for the schooling species. It does not protect the individual under attack but operates to the benefit of the group" (Hoar, 1966, p. 469). As Pfeiffer (1962) concludes, the fright reaction may well have played a profound role in the evolution and overwhelming success of the Ostariophysi in freshwater habitats.

Some four fifths of the 5,000-odd species of Ostariophysi are either cyprinids or catfishes. Most of the remaining species are characids (Characidae), which have undergone a remarkable adaptive radiation. "For example, there are small pelagic, herringlike forms (genus *Thrissobrycon*); toothless mullet-like forms (genus *Curimatus* and relatives); elongate gar-like characids (genera *Boulengerella* and *Ctenolucius*); characids similar to members of the cyprinid genus *Labeo* (genera *Ichthyoelephas* and *Prochilodus*) and darter or goby-like forms (genus *Characidium*). The genus *Salminus* in many respects resembles the trouts and salmons: the characid genus *Grundulus* has been mistaken for a cyprinodont fish; *Bivibranchia* would probably be mistaken by the uninitiated for the marine *Albula* of the order Isospondyli. Many of the small characids of the genera *Hemigrammus*, *Aphyocharax*, *Bryconamericus*, and other close relatives parallel in habitat and body shape the divergent habitats and morphological forms of many of the Asian, African, and North American minnows . . . In addition, the characids have developed many morphological types that are adapted to ecological habitats unique among fishes—for example, among them are found the only true flying fishes, which utilize their pectoral fins for propulsive force while in the air (genera *Thoracocharax*, *Gasteropelecus* and *Carnegiella*). Among the characids are found the piranhas (genus *Serrasalmus*), the deep-bodied, voracious and much feared predators of South American rivers. Strange to say, some of the close relatives of *Serrasalmus* (genera *Myleus* and *Metynnis*) limit their diet mainly to fruits and other vegetable materials" (Weitzman, 1962). But what characters of the characins seem to have contributed outstandingly to such success? All one can say, even after scrutiny of Weitzman's (1962) thorough study of their osteology, is that the ancestral characid must have possessed a unique set of potentially adaptable features.

Certain characters involved in the success of cyprinids and catfishes are more apparent. Compared to a generalized characid, the cyprinids are specialized in their protrusile and edentulous jaws. The sudden increase in buccal volume as the jaws are shot forward is correlated with a well developed pharyngeal dentition. Relatively small food organisms are thus inhaled and passed to the gripping, stoking teeth in the throat. By these mechanisms, a great many cyprinids take full advantage of the abundance of insect larvae in freshwaters. Others pump in mud and strain out algae and small animals through the pharyngeal teeth and gill rakers. There are also types of herbivorous and predatory cyprinids. But as Brittan (1961) contends, the cyprinids are not "over-specialised" fishes, which may well have "enabled them to move, or begin to move, into every freshwater ecological habitat (except mountain torrents and markedly oxygen-deficient waters)."

Catfishes show the greatest adaptations to stagnant conditions. Regan (1936, p. 230) described them as "carnivorous fishes that live at the bottom, many of them in stagnant water or muddy rivers; the eyes are often small, but these fishes find food by means of their barbels, which are generally six or eight in number, and may be very long." Catfishes indeed fit into ecological gaps—in time as well as space—left by the cyprinids and characids. Catfishes not only tend to have small eyes but to be nocturnal, and they depend on elaborate olfactory organs besides the gustatory and tactile endings on their barbels (and on other parts of the body) for finding their food. Most of the characids and cyprinids have large eyes and diurnal habits. Moreover, in conformity with their bottom-dwelling tendencies catfishes have a rather reduced swimbladder and a depressed body form (Alexander, 1965). Alexander also draws attention to the protection given by their spines, which in most species are in the dorsal and pectoral fins. Through the above developments and others, catfishes have become masters of many freshwater habitats.

In ubiquity, diversity, and often in numbers of individuals, spiny-finned teleosts are dominant in shelf and coral seas: in the deep ocean they are much less diverse than soft-finned groups (for example, stomiatoids, myctophids, macrourids, and brotulids). If the mail-cheeked fishes (Scleroparei) are placed in the Percomorphi, some 8,000 species are comprised. Thus, at least half the kinds of modern marine fishes are percomorphs. Moreover, successful groups such as the flatfishes and plectognaths had a percomorph ancestry.

Percomorph fishes are both simpler and more complex than the more primitive, soft-rayed teleosts. They have lost the serial array of intermuscular bones, which may be correlated with a reduction in the number of vertebrae. Typically, they develop seventeen principal caudal rays instead of nineteen, and their skeleton lacks bone cells. And while the more primitive soft-rayed and spiny-rayed teleosts have labyrinthine outlets of their head lateralis canals, this system tends to be simpler in most of the percomorphs.

The more advanced and complex characters of the percomorphs are these: a closed (euphysoclistous) swimbladder, capable of rapid, as well as longer term, control of buoyancy; flexible fin patterns coupled with a body form allowing of fluid maneuverability; spiny armament; mobile jaws and pharyngeal tooth-plates; and lastly, an apt kind of visual cell mosaic in their retinae. At the very least these advanced features are behind the complex adaptive radiations of percomorph fishes.

Spiny-finned teleosts are not alone in developing a euphysoclist swimbladder (Fänge, 1953; Marshall, 1960), but their organization is uniquely correlated with an outstanding attainment of physoclists—their fine control of buoyancy by swimbladder regulation "with a sensitivity and precision comparable to postural regulation in terrestrial mammals" (McCutcheon, 1966). McCutcheon has provided more evidence that physoclists (and members of the Ostariophysi) control the effects of slight changes in pressure by muscular compression of the swimbladder. Certain physoclists, for instance, the percoids *Lagodon rhomboides* and *Centropristes striatus*, reacted to changes of less than 0.5 cm H_2O in pressure, and in a time of 0.1 second. In a sine tube, these and other physoclists responded to pressure-induced depth changes by movements that kept the fishes, or brought them back, to a reference position fixed by their eyes. After pressure changes from 0.1 to 1.5 cm H_2O, position was often kept through eye scanning and apt pectoral activity so that the pelvic fins just touched ground. After greater changes in pressure cyclic swimming away from the fixed position was followed by movements against the buoyancy trend. Again, the eyes scanned and the paired fins were used together, this time as the fish touched ground to test the pressure gradient. In nature, comparable activity may well be vital as tidal levels change. McCutcheon also suggests that "the known territorial defence activity, nest protection activity, hiding behaviour, and protective coloration and camouflage structures for evasion of predators emphasise the importance of buoyancy regulation to hold a position." Percomorph fishes are certainly masters at using cover, and in keeping station. Again, one has only to think of a coral reef.

Quick and precise control of buoyancy is thus correlated with versatile action of the paired fins, which, *inter alia*, hold the fish steady enough for visual scanning. As already stated, in spiny-finned fishes the laterally set pectorals have a near vertical axis, and they are inserted near the center of gravity. Thus, pectorals can become pivots, oars, or brakes, in concert with the nearby pelvics (Harris, 1953). Moreover, the evolution of pectoral pivots in percomorphs evidently entailed a shortening of the trunk and some deepening and compression of the body (Patterson, 1964). In fact, the fin patterns and body form of typical percomorphs make for great maneuverability. By turning quickly, prey may be surprised and enemies avoided. Fine control of movement, posture (and buoyancy) is invaluable in many other ways. Consider the neat adjustments of position that are part of the many feeding skills of diverse percomorph fishes in coral reefs (Hiatt and Strasburg, 1960). Feeding skills also depend on protractile jaws. What has been said about the cyprinids applies equally well to many percomorphs. Moreover, many species feed, at least in part, on bottom-dwelling organisms. These fishes, as Alexander (1967b) suggests, may benefit from their protractile jaws in that "it is an advantage, when feeding from the bottom, to have the axis of the body as nearly horizontal as possible. A fish must return to the horizontal before it can flee from a predator." One is reminded of the surgeon-fishes and siganids, which have fixed jaws and stand on their heads when browsing on plants. But to compensate, as it were, for their vulnerable feeding postures, they bear formidable spines on the caudal peduncle. It is striking also, that the adaptive radiations of isospondylous fishes, very few of which develop protractile jaws, have produced only a handful of benthic species.

Once the prey is seized, spiny-finned fishes pass it to the pharyngeal teeth. Concerning the mobility of these teeth, much can be learned from a study of the branchial muscles (Nelson, 1967). In particular, the development of paired muscles both ventrally (rectus communis) and dorsally (retractores dorsales), which attach to the toothed pharyngeal plates, is a considerable advance on the more generalized complex of branchial muscles found in lower teleosts, such as *Elops* and *Aulopus*. The spiny-finned fishes thus have more mobile pharyngeal "jaws," which have evidently taken over much of the function of the outer jaws in the seizing and manipulating of food. Nelson concludes that "this loss in function may account for the subsequent radiation of jaw types which has occurred among higher teleosts (for example, those in Gregory, 1933). This, and the contemporary radiation in gill arches, seem to be the morphological basis for many of the specialized food habits of the higher teleosts, and therefore may account for their dominant position today among fishes."

With regard to visual aspects, there are two main cone patterns in the retina of teleosts: double cones arranged in parallel rows (sometimes alternating with rows of single cones) and a mosaic formed of square units. In fishes not dependent on sharp vision, such as diverse benthic and nocturnal forms, there is virtually no cone pattern in the retina. Rows are typical of cyprinids and gadids, and squares "are consistently present within many of the families of the Perciformes" (Engström, 1963). Rows are also found in salmonids and clupeids, but in the herring and sprat, there is a square mosaic in the lower part of the retina, which is adapted for the acute vision needed to detect small planktonic animals. The pattern giving good acuity is thus typical of percomorphs. Details of their living space, food, members of their own kind, and so forth, are presumably clearly registered by their visual systems, which fits well with earlier remarks on the importance of vision in percomorph behavior.

Such then, are some of the factors behind the dominance of teleost fishes. We have measured this mastery

The Success of Teleostean Fishes

by the numbers of individuals, species, and living spaces. We should also remember that the total biomass of teleosts greatly exceeds that of any other group of living fishes. By these criteria the elasmobranchs are not successful fishes. Even so, they have maintained their place in the life of the ocean for many millions of years. But how far could they have expanded their place had they developed a swimbladder?

2 Features of Dynamic Design

Fishes swim and breathe in a medium that is eight hundred times denser than air. They oscillate fin and body surfaces across or along their line of motion, so exerting a series of backward thrusts. In a given time the energy used in swimming is essentially the product of the fishes' speed and the resistance it encounters. For fish-sized animals most of this resistance is due to the inertia of water, which must be shouldered aside and accelerated backwards. If the water flowing around the fish approaches streamlined patterns, the resistance and speed will be close to their optimal levels for a given expenditure of energy.

Studies in fluid dynamics show that a streamlined body is fusiform. The deepest part of the body lies close to a point one third of the way along the major axis. Evolutionary processes have produced a wealth of fishes with bodies more or less resembling this model shape. The main difference is that fishes tend to be elliptical rather than circular in cross section (Figure 2). But such a body form gives more lateral surface for exerting thrust, which is integral to the locomotion of most fishes. Fusiform fishes that come closest to being circular in cross-section are the tunas and the isurid sharks. These fishes virtually depend on rapid oscillations of their caudal fin for locomotive thrust. The caudal peduncle, which in most spindle-shaped fishes is compressed and flat-sided enough to provide some thrust, is rounded and fitted with streamlined lateral keels in the tunas and isurid sharks. Their motive power being so concentrated on the tail fin, it is not altogether surprising that the body form of tunas is close to that of aircraft with a laminar fuselage (see Hertel, 1962, and Figure 3). It seems reasonable too that squid, whose quick movements stem from a rapid series of jets from the mantle cavity, should have a body that is circular in cross-section. But the fastest swimming squid, such as *Dosidicus*, have an overall shape that is slimmer than the theoretical laminar form of hydrodynamics (see Figure 35).

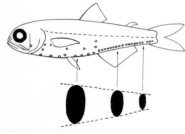

2 *Electrona subaspera,* a lantern-fish showing (in solid black) cross-sections through the flat-sided flanks.

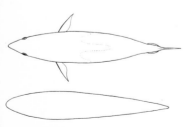

3 The skipjack (*Katsuwonus pelamis*) seen from above. Below: a streamlined body.

22

Features of Dynamic Design

There is convergence between fish shapes and those of other aquatic vertebrates; ichthyosaurs, penguins, cetaceans, and seals. This is not surprising in so inert a medium as water. Few land-dwelling animals are streamlined; notable exceptions are various swift-flying birds. But if the thematic shape of many aquatic animals is fusiform, fishes have played the most variations on this theme. One has only to think of the fish faunas of coral reefs or of the mid-waters in the deep ocean.

A fish is swaddled and shaped by criss-crossing geodesics of fibre tracts and scale rows in the skin (Marshall, 1965b). The skin is also attached to the axial muscles, which in active fishes make up 40 to 65 percent of the body weight. But to escape quickly from enemies, capture prey, and so forth, demands large muscles when the medium to be moved aside and accelerated is so heavy. Moreover, the amount of oxygen dissolved in water, even when close to saturation, is about 30 times less than the amount in air. To supply massive muscles and other organs with oxygen requires a highly developed gill system. Large, efficient pumps to move the heavy medium over the gills, a high surface area of gill per unit weight of fish, and countercurrent flow of blood and water across the gill lamellae are all necessary attributes of active fishes. When one considers also that jaws, mouth, brain, sense organs, and so on, must all be accommodated in the head, and that the head, at least in active fishes, should move water aside without undue disturbance, one realizes how superbly shaped and designed is this part of the organism.

Close study has shown that fusiform fishes swim by means of body waves that sweep back with increasing amplitude (Gray, 1968). Such waves come from the dynamic interplay of myotomes, myosepta, and vertebrae. Now, if fishes are to gain a good purchase on their medium, their bodies must undulate in smoothly flowing curves. This is ensured by the zigzag nesting of the myotomes, which means that the number of contracted fibres in any given transverse plane gradually increases until all the fibres in that plane have contracted; thereafter, the number gradually diminishes. In this way a wave moves easily and evenly down the body. Moreover, the thickening of the muscle fibres during contraction can only be contained if there is considerably less than a right angle between the myosepta and the axes of the fibres: hence, the oblique setting of the myotomes (Willemse, 1966).

Through the rapid side-to-side switching of muscular waves, the tail and its fin are sculled across the line of motion. During these oscillations, body and fin surfaces with some backward component of motion provide the thrust. The magnitude of thrust of such a surface depends directly on: (1) its area; (2) transverse speed; (3) extent of swing; and (4) angle of attack.[1] Compared to a body surface of the same area, a flexible tail fin thus has the advantage in all thrust factors. The tail fin of fusiform fishes should thus exert the main thrust, but the proportion of this to the total thrust varies from one species to another. Concerning this aspect, Bainbridge (1963), who made relative assessments of factors 1 to 4, has compared the propulsion of dace (*Leuciscus leuciscus*), goldfish (*Carassius auratus*), and bream (*Abramis brama*). The percentage of the total thrust contributed by the caudal fin is 84 in the dace, 65 in the goldfish, and 45 in the bream. These differing figures are related, as Bainbridge says, to differences in shape, skeletal structure, and body musculature. Considering shape alone, the relatively low figure for the bream is understandable in view of its long and deep caudal region. The dace also has a fairly long tail but it is slimmer and more rounded. In fact, the more streamlined the fish, the greater the relative thrust of the caudal fin. We realize too that the deep-bodied bream lives in quiet waters, whereas the dace faces lively streams.

If we continue the series bream–goldfish–dace, we should end with the scombrid fishes. The more specialized kinds, such as the blue-fin tuna and the skipjack, have a body shape more like an ideal streamlined form than has a dace (Figure 3). The tuna tapers strongly and elegantly to a slim, but strong, caudal peduncle, which bears on each side a keel. Body waves are much less evident in the swimming of a tuna; the only part that bends to any extent is the caudal peduncle. Moreover, the bases of the strongly built caudal rays straddle the hypural bones, so making the fin much more rigid than the tail of a dace (and of most other fishes).

Tunas have a lunate tail fin of high aspect ratio, which they beat very rapidly. At first sight one might imagine that the angles of attack of such a rigid fin would tend to be low. Indeed, high aspect-ratio aerofoils generate a high lift force at low angles of attack. But recent studies

1. A propelling surface will also be most effective in undisturbed water, which is rare in many parts of the hydrosphere.

Features of Dynamic Design

by Fierstine and Walters (1968) do not support a prediction from aerofoil to hydrofoil. A tuna's tail is double-jointed, the vertebral structure being such that one joint is just ahead of the first vertebra of the caudal peduncle, the other just behind the last vertebra. After the last peduncular vertebra come the three last vertebrae of the column, which are very compressed and serve to support the skeleton of the caudal fin. The peduncular vertebrae themselves form a rigid unit splayed outward to form a bony keel on each side (Figure 4).

Tendons of the swimming muscles, which insert on the bases of the middle caudal rays, run just above and below each bony keel. The keels act as a pulley, the angled run of the tendons being such that their pull is more powerful than it would be if keels were not present. The tendons are composite structures, formed from the outer tendon sets that arise from the anterior cones of the muscle segments (Figure 4). Moreover, these tendons transmit the pull generated by the white fibres of the myotomes, which are used during rapid swimming. Sustained pull for cruising comes from the red muscles and is transmitted through separate (posterior oblique) tendons to all the vertebrae forward of the second peduncular vertebra.

Fierstine and Walters studied the swimming movements of wavyback skipjack (*Euthynnus affinis*), which were filmed as they cruised around a circular tank, 23 feet in diameter. Analysis of the film showed that the transverse velocity of the caudal fin increases from zero at the extremities of its swing to a maximal value close to the line of motion. As the fin crossed this line, the angle of attack averaged 32.4° for sequences from five fishes and ranged from 20° to 60°. For all positions the mean angle was 29.2°, but values as high as 100° were measured (Figure 4).

Amputation tests on two yellow-fin tuna (*Thunnus albacares*) indicated that the caudal rays contribute at least 90 percent of the forward thrust in swimming. The high angles of attack, which contrast with the 11° angle of an efficient aerofoil, are related to the double-jointing of the caudal peduncle. Bainbridge (1963) estimated mean values of 13–14° for the dace and goldfish. Like most kinds of fusiform fishes these two species do not have a dual peduncular joint but their caudal rays are more flexible and mobile than those of scombroid fishes.

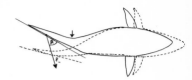

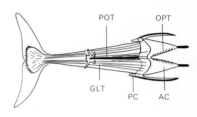

4 Top: Way of measuring angle of attack (α) of a tuna tail fin taken from two successive ciné-frames. The arrow against the solid-outline figure shows the position of the prepeduncular joint. Middle: a scombroid muscle segment and its tendons. *AC* and *PC*, anterior and posterior cone of segments; *OPT*, origin of posterior oblique tendon; *POT*, posterior oblique tendon; *GLT*, great lateral tendon. Bottom: caudal vertebrae complex of a black skipjack; the two prepeduncular joints are marked by arrows. (Redrawn from Fierstine and Walters, 1968.)

The unique features of scombroid fishes are not only this dual joint but also the elaborate and massive system of tendons developed from the muscle segments. Swordfish (Xiphiidae), marlins, spearfishes, and sailfishes (Istiophoridae), which have been classified as scombroids, are without these features. These, and other differences have led Fierstine and Walters to the reasonable conclusion that the xiphiids and istiophorids should be placed in a suborder (Xiphioidei) apart from the scombroids.

Observations on the locomotion of the larger scombroid fishes are not readily made. Hooked yellow-fin tuna (*Thunnus albacares*) and wahoo (*Acanthocybium solandri*), both about a meter in length, swam with a scaled line at over 2,000 cm/sec., about 40 knots (Walters and Fierstein, 1964). Studies of the Pacific bonito (*Sarda chiliensis*) in captivity showed that cruising fish moved at 88 cm/sec. at a mean tail beat frequency of 1.42 beats/sec., (1 knot = 51.48 cm/sec.). Magnuson and Prescott (1966) also observed that the most rapid speed was 370 cm/sec. At four tail beats per second, 57 cm bonito swam at 170 cm/sec., 57 cm skipjack (*Katsuwonus pelamis*) at 230 cm/sec., and 52 cm. yellow-fin tuna at 240 cm/sec.

Considering just the figures for the skipjack and bonito, for they not only relate to the same frequency of tail beat but also to individuals of the same length, how is the skipjack able to cruise a knot faster? Both fishes presumably have the advantages of a scombroid mode of swimming. But the tail fin of the skipjack has a much higher aspect ratio (7.49) than that of the bonito (5.71).[2] This may give the skipjack an advantage. Moreover, length for length the skipjack is heavier than a Pacific bonito. The skipjacks' chunky body has closely knit muscles, which are associated with a firm skeleton; the skeleton of a bonito is spongy and rather weak, and the flesh is rather soft. In cross-sections of the body at comparable levels, there are more myotome rings and greater sectional area of red muscle in a skipjack. Since the red muscles are used in cruising, the skipjack should be able to put more power into each tail stroke than the bonito. Moreover, the temperature of the red muscles of the skipjack will be higher than that in bonito red muscle, which is without a countercurrent retial system for the conser-

2. Aspect ratio is span of tail fin^2/area, and the figures are taken from Fierstine and Walters (1968). (Fierstein was changed to Fierstine and the distinction has been preserved here.)

vation of heat. Lastly, the body of the skipjack is closer to an ideal laminar form than that of the bonito.

When Bainbridge's (1963) formula, ($V = \frac{1}{4}(L(3f - 4)$, where V is the speed in cm/sec., L the length of the fish in centimeters and f the tail beats/sec, is applied to Walter's and Fierstein's (1964) observations on the speed of hooked yellow-fin tuna and wahoo, the tail beat frequency comes to nearly 30 beats/sec. This frequency seems very high for such large fishes, but there is no doubt that when the larger scombroids travel at speed the tail is oscillated rapidly. Considering the yellow-fin, high muscle temperature should be a help. The temperature of the white muscle, which is used in sprints, may be 2 to 10°C higher than that of the sea, and with a rise of 10°C muscular contraction and relaxation become three times faster (see Carey and Teal, 1966b). The limitation to the frequency of beat is, of course, ultimately set by the dynamic form of the caudal peduncle and tail fin. The former structure is slim and keeled, and the latter has a high aspect ratio, both favoring a high rate of beat. But the wahoo, which is without heat-conserving retia, has a stouter peduncle than the yellow-fin and the caudal has a lower aspect ratio (6.3 as compared to 7.69). Clearly, there is more to be learned.

One disadvantage of the high aspect ratio caudal fin of scombroids and other fishes is its rigidity. Fishes with soft caudals, formed of individually movable rays, can alter the area of the fin during its lateral sweep and so smoothly control the thrust.[3] A tuna, which has no such caudal smoothing, may well be poorly maneuverable at low speeds (Bainbridge, 1963). But this is not to say that tuna and their like are not able to make quick turns. After thunniform "caudal keels" were fitted before the screw of merchant ships, the turning circle was halved (Watts, 1961).

If the body waves of spindle-shaped fishes are regarded as the middle part of the spectrum and those of thunniform fishes as one extreme, the other extreme is provided by eel-shaped species. In such fishes the caudal fin is so reduced and the form so serpentine that all the thrust comes from waves of high amplitude that flow down the body. Again, the magnitude of thrust exerted by a

---

3. The fin rays are moved by a complex of intrinsic muscles, which are reduced in tunas.

particular body surface depends on its speed of lateral sweep and the angle of attack (Gray, 1933). Between anguilliform and fusiform parts of the spectrum there are many fishes with "subanguilliform" body waves. Dogfishes are good examples.

When not feeding, courting, avoiding enemies, and so forth, active fishes tend to move at their cruising speed. For 15 cm dace, trout, and goldfish this speed is about 3–4 body lengths per second. Over intervals of 1 second and 5 seconds they can sustain speeds of 10–12 and 6–10 lengths per second in that order. As Bainbridge (1960) concluded: "The ability to sustain high speeds is thus much lower than that suspected previously."

At their cruising speeds, the flow of water around fusiform fishes is likely to be nearly laminar. What happens when they sprint? Little is known, but at all events, they can accelerate quickly. Starting from rest, species such as rainbow trout (*Salmo gairdneri*), pike (*Esox lucius*), carp (*Cyprinus carpio*), rudd (*Scardinius erythrophthalmus*), and dace (*Leuciscus leuciscus*), which ranged in length from 13.5 to 22 cm covered a distance of 5 cm in $\frac{1}{20}$ second, thus accelerating at 40m/sec./sec. (Gray, 1953). This seems remarkable considering the inertia of water.

In this medium, quick, coordinated flexures of the body are needed if a fish is to escape from an enemy or surprise its prey. Part of this fitness for life in water is centered in the Mauthnerian system, which consists of a pair of large neurones and giant axons. The neurons (up to 200μ in length) lie near the middle of the medulla opposite the roots of the vestibular nerves. The axons decussate close to their origin, descend the entire length of the spinal cord, and at intervals send off collateral branches to primary motor neurones (Figure 5). Synaptic endings on the two dendrites of each Mauthner cell link them with parts of the acoustico-lateralis, optic, and cerebellar systems. In brief, the Mauthnerian apparatus short-circuits the reflex paths between sensory centers and motor nuclei of the central nervous system. This parsimony in synapses, coupled with the large diameter and heavy myelin sheathing of the Mauthner fibres, provides for the sudden and coordinated response of the swimming muscles (see also Yasargil and Diamond, 1968).

Knowledge of the forms, activity, and ecology of fishes fits well with the idea that the Mauthnerian system is essentially a device for rapid escape. This system is not only

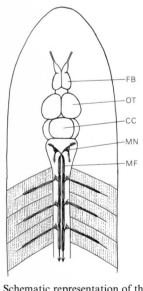

5 Schematic representation of the Mauthnerian system. *FB*, forebrain; *OT* optic tectum; *CC* corpus cerebellum; *MN*, Mauthner's neurone; *MF*, Mauthner's fiber.

developed in fishes but also in certain amphibians.[4] In fishes it is present in lampreys (ammocoetes), chimaeras, and bony fishes, but not in sharks and rays. Considering the teleosts first, the Mauthnerian system is well formed in many species but is reduced in eel-shaped and benthic forms. In certain benthic species De Angelis (1950) found a reduced system in a weever (*Trachinus radiatus*), *Gobius jozo*, two blennioids (*Clinus argentatus* and *Blennius basilicus*), *Scorpaena porcus*, and *Solea impar*. Both the neurones and axons are small, the latter being no larger than most of the fibres in the spinal cord. But such fishes and other benthic species, which have no swimbladder and habitually rest on the bottom, are not primarily dependent on caudal oscillations for propulsion. They rely on concealment and camouflage rather than sudden flight for evading their enemies.[5] In eels the reduction of the Mauthnerian system may well be related to their form. A spindle-shaped fish can suddenly dart away because the more massive front half of the body acts as a pivot for the oscillating tail. No such bracing is possible in eel-shaped fish.[6] But the ribbon-like deal-fish, evidently has a well-developed Mauthnerian system. Here the flattened forward parts of the fish presumably form a pivot for rapid undulations down the rear end of the body.

Of all fishes, no such pivot is lacking in the scombroids, but the larger species, at least, have a reduced Mauthnerian system. Instead, according to Stefanelli (1962), they have a longitudinal median fasciculus, which extends from mesorhomboencephalic centers of the tegmentum. Stefanelli argues that a Mauthnerian system is not sufficient in these large, fast moving fishes. But is the regression of this system related to the form of the caudal fin? Tuna and their kind are always on the move, and even supposing "a sudden burst of speed from a standing start is required, the angle of attack of (their) kind of fin would be too high and the tail would 'stall'" (Harris, 1953). We should also consider the warm-blooded muscles of tunas. The warmth of the muscles will be taken up

4. In tadpoles and in aquatic urodeles with well-developed tails.

5. These secretive, lurking kinds of teleosts make great use of their pectorals as oars. So do the more active, conspicuous wrasses, which have a well-developed Mauthnerian system. But wrasses, if startled, are well able to use their tails to make a dart.

6. The Mauthnerian apparatus is reduced in the sea lamprey but is well developed in the ammocoete, which lives in a burrow and down which it may need to withdraw suddenly.

by the motor nerves and so speed up the transmission of nerve impulses. Muscles near the vertebral column may be 10°C warmer than the sea, and this heat is presumably absorbed by the spinal cord. Thus, transmission down the longitudinal fascicular tract should also be quickened. Perhaps these are the main reasons why tunas are able to do without Mauthner cells and giant axons.

It is not surprising that the rays (Batoidei), which are bottom-dwellers, or derived from such, and do not have the form or means for sudden acceleration, lack a Mauthnerian system. Concerning the absence of this system in sharks, Aronson (1963) notes that, while some sharks swim in anguilliform fashion, others are more carangiform in style. He adds: "Most sharks are constantly in motion. In fact, this is necessary in many sharks as an aid to respiration. Sudden spurts in forward locomotion are rarely seen. The absence of the Mauthner apparatus in sharks may be the explanation for their inability to accelerate rapidly." Did the ancestors of sharks and rays rely more on armor than evasive action for keeping their lives?

Teleosts with moveable caudal rays are able to change the area of the fin during its lateral sweeps. Bainbridge (1963) also found that there was virtually no instant when the fin was not exerting a thrust. In particular the central part of the fin, even when passing the mid-line, could still have a positive angle of attack. Moreover, the upper edge of the caudal is not always vertically above the lower edge. For instance, at the end of a lateral stroke, the upper edge may lie behind the lower edge but as the fin moves back to the middle of its sweep the upper edge catches up with the lower.

The dynamic consequences of this caudal twist are doubtless rather complex, but one can see that resistance to lateral motion will be least when the fin is most turned away from the vertical plane—and is also just starting its return stroke. At such times the angle of attack of the lagging upper lobe of the fin will be greater than that of the lower; so the overall thrust will be more than would occur without this lag. At all events, it is clear that caudal twisting is an integral part of the propulsive cycle, which may also be appreciated by study of the structure of the caudal fin. Caudal fin twisting must be effected by the longitudinal hypochordal muscles. When well developed, each muscle is (typically) inserted on the hypural bone below

Features of Dynamic Design

the terminal caudal vertebra and on one or more suc-
ceeding hypurals (Grenholm, 1923; Nursall, 1963a). On
each proximal side of the terminal hypural is a process
that Nursall calls the hypurapophysis, which, he says, is
an attachment for the forward and lateral parts of the
hypochordal muscles. From their hypural attachments
these muscles extend obliquely upward to end in tendons that insert behind the hinges of the uppermost (usually 3 to 5) principal caudal rays (Figure 6). Contractions
of either hypochordal muscle must pull these rays side-
ways and downwards. The switching of contractions
from one side to the other thus gives the fin twisting described above.

Study of Grenholm (1923), Ford (1937), Nursall
(1963a and b), and a series of skeletons in the British
Museum shows that hypurapophyses are well developed
in nearly all teleosts that habitually swim by caudal oscillations. Notable exceptions are the gadoids, which have a
reduced caudal complex (*sensu stricto*) and the Heterosomata (Grenholm, 1923, and personal observation). In
adults of both groups the hypochordal muscles are absent. Now, flatfishes swim by undulations of their compressed bodies and it is likely that the caudal *region*
provides much more thrust than the caudal *fin*. But in
pre-metamorphosed turbot, *Scophthalmus maximus*,
which swim in normal fish-fashion and have a relatively
large caudal fin, the hypochordal muscles are well developed. They remain as rudiments in the adults (Grenholm, 1923). It looks, then, as though the twisting
mechanism of the caudal fin is prominent when the fin is
fully developed and is providing much of the thrust. But
what of fishes that are not altogether dependent on caudal
locomotion? For instance, surgeon-fishes (Acanthuri-
dae) often use their pectoral fins as oars, while wrasses
(Labridae) and parrot-fishes (Scaridae) rely even more
on pectoral propulsion. It is thus interesting to find that
the hypurapophysis and hypochordal muscles are well
developed in surgeon-fishes, whereas in wrasses and
parrot-fishes these processes are very small or absent and
the muscles small (Figure 6). Cottids, nototheniids and
gobies, which live on the bottom and make great use of
their pectorals as oars, also have small processes and
muscles. But all such fishes use the tail and its fin when
hard pressed. Perhaps escape will be quicker if some
means are retained for twisting the caudal member.

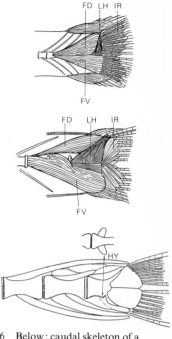

6 Below: caudal skeleton of a
barracuda, showing the position of
the hypurapophysis (*HY*). Middle:
part of the tail musculature of a
barracuda and a wrasse (*Labrus ossi-
fragus*), above. Note the much larger
longitudinal hypochordal muscle (*LH*)
in the former. *FD*, flexor dorsalis; *FV*,
flexor ventralis; *IR*, interradial
muscle. (Redrawn from Grenholm,
1923.)

The complement of main caudal rays is also related to the mode of locomotion. In the orders Isospondyli to Berycomorphi (Regan's 1929 classification), species with a fully developed caudal fin usually have 19 principal rays and the caudal region is the prime means of propulsion. Comparable species of the Percomorphi and related orders usually have 17 principal rays. But fishes that cruise by other than caudal means of propulsion most often have a reduced number of caudal rays. Sea horses and pipe fishes, which not only move by rapid undulations of their dorsal and pectoral fins but have evolved a stiff armor, have an extremely reduced caudal fin. The plectognath fishes, with heavily armored or stiff bodies, are driven by motions of their opposed dorsal and anal fins, aided by the pectorals (tetraodontids, diodontids) or the caudal fin (ostraciontids). In all members of this order the caudal fin has a low aspect ratio and a reduced number of rays. Tyler (1962) counts 12 in triacanthoids and balistoids, 10 or 11 in ostraciontoids, 11 in tetraodontoids, 9 in diodontoids, and none in most molids. The Zeomorphi are another entire order with a reduced caudal complement (12–15 rays). All members probably swim, like *Zeus* and *Capros*, by means of undulations down their opposed dorsal and anal fins (Figure 7).

Surgeon-fishes which often use their pectorals as oars, have a slightly reduced caudal complement (16) compared to other percomorphs. Wrasses and parrot-fishes, depending much more on pectoral locomotion, have fewer principal caudal rays (usually 12–14).

All the foregoing fishes with a reduced caudal have a swimbladder, and they swim easily in aquatic space. But this organ is absent in true bottom-dwelling fishes, some of which make great use of their fan-like pectorals for propulsion. Again, such fishes have a reduced number of main caudal rays. For instance, most nototheniiform fishes have 14 caudal rays. A reduced caudal fin is present also in scorpaenid, cottid, and blennioid fishes.

In all of these fishes, particularly the free-swimming kinds, the caudal fin is used more often as a rudder than a propellor. In keeping with this change in functional emphasis the caudal has come to have fewer main rays and a low aspect ratio, but it is still well enough formed to act as a propellor if need be.[7]

7 Two fishes with a low aspect-ratio caudal fin and a reduced number of caudal fin rays. The soft dorsal and anal rays, the principal means of locomotion, are shown black. Above: John Dory (*Zeus faber*). Below: a filefish (*Amanses sandwichensis*).

7. Outstanding exceptions are the flying-fishes, half-beaks, garfishes, and sauries, which have 13–15 main caudal rays.

Features of Dynamic Design

There is a further functional correlation in fishes that use their pectorals as paddles, Dijkgraaf (1962), following Plate (1924) and Popovici (1931), contends that when these fins beat towards the flanks "The resulting currents would hit the anterior part of the main lateral line of the trunk and hamper its functioning—if it were still in the "normal" position. However, in fishes of this type, the relevant part of the trunk line is displaced toward the dorsal side, thus avoiding the area of disturbance." Such displacement is well displayed, for instance, by noto-theniiform fishes, wrasses, parrot-fishes, and various blennioids. Indeed, a general correlation between the positions of the pectoral fin and the lateral line is evident among teleosts.

In most members of the orders Isospondyli and Os-tariophysi, the pectoral fins emerge low down on the shoulders, well below the median longitudinal axis of the body. Typically, the lateral line runs close above this axis in the forward half of the body but slopes gently down so that the line marks the median axis over the backward half. In groups where the pectorals have a higher lateral setting, so that their bases cover, or are not far below, the median axis, the forward part of the lateral line is deflected about midway between the body axis and the dorsal profile (Figure 8). Indeed, in the deeper-bodied percomorph fishes the lateral line may only assume a median setting along the caudal peduncle. Apart from the perco-morphs, other groups with lateral pectorals and dorsally displaced lateral line include the myctophoids, gadoids, macrourids, zeomorphs, Scleroparei, and plectognaths.

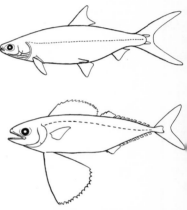

8 Position of pectoral fins and lateral line. Above: pectorals and line low (*Chanos*), as in "lower" teleosts. Below: pectorals and line high (*Gasterochisma*), as in "higher" teleosts.

Breder (1926) and Harris (1953) have shown that pectorals with a near-vertical axis and a lateral setting are broadly adaptable: they may function as oars, brakes, pivots, and if need be, as back-paddles to offset the impulsion of water out of the gill chambers. If the forward part of the lateral line ran medially, such use of the pectoral fins would either mask or "confuse" the lateral line. There is no such danger in fishes with low-slung pectorals, which tend to function as hydrofoils and balancers. In passing, we should remember also that the lateral line of surface-feeding fishes is generally deflected toward the underparts of the body.

Sharks are classic instances of the use of pectoral fins as hydroplanes. The lift from their cambered pectorals, coupled with that generated during the lateral move-

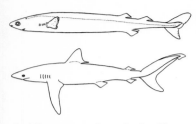

9   Fin patterns of two sharks. Above: *Isistius brasiliensis,* a small, neutrally buoyant, deep-sea species, with paddle-like pectoral fins and nearly symmetrical caudal fin. Below: the silky shark (*Carcharinus floridanus*), negatively buoyant, with hydroplane-like pectorals and a markedly asymmetrical caudal fin.

ments of their heterocercal tail fin, offsets their negative buoyancy. But this is not true of all sharks. The small mesopelagic shark, *Isistius brasiliensis,* as I found, is close to being neutrally buoyant. Since there is little or no need for uplift, it is surely significant that the pectoral fins are not shaped and set as hydroplanes: they are paddle-like and placed laterally on the shoulders. Moreover, the upper lobe of the caudal fin is about equal in area to that of the lower lobe, not very much smaller, as in sharks with a heterocercal caudal (Figure 9). Like neutrally buoyant teleosts, *Isistius* has been free to evolve paddle-like pectorals and a symmetrical tail fin. So have two other mesopelagic sharks, *Euprotomicrus* and *Squaliolus,* which are also, one presumes, near to being neutrally buoyant.

3 The Life of Deep-Sea Fishes

Deep-sea fishes live in oceanic waters where sunlight is perpetually faint or lacking. Even so, the surroundings of many species are not always so restricted. Diverse meso-pelagic fishes of the twilight zone spend both their larval life and much of their adult, "night" life in the plant-bearing waters near the surface. Most kinds of bathypela-gic fishes from the sunless depths pass no more than their early life in the surface waters. Though we know much less of the entire life of the bottom-dwellers, many species may well never depend on a near-surface nursery ground (Marshall, 1965a). At the very least, then, deep-sea fishes pass their adult existence under the cover of twilight or darkness. Yet this cover is lifted here and there by flashes of living light.

In clear oceanic waters the sun's rays are strong enough for photosynthesis down to a level of about 100 meters. Between depths of 150 to 1,200 meters, again in very transparent waters, is the twilight zone, where plants, if they can live at all, must be heterotrophic. The twilight zone merges below with sunless waters, which thus form the greatest part of the deep ocean. If twilight becomes darkness at a mean depth of 1,000 meters, and taking the mean depth of the open ocean as 4,000 meters, the percentage volumes of the euphotic zone, twilight zone, and sunless zone are respectively, 2.5, 22.5, and 75.0 (Figure 10). But this is true only of the subtropical and tropical regions, though these are the headquarters of deep-sea fishes. In temperate and polar waters the threshold of light has a lower mean depth. Except, it seems, for the deepest reaches of the ocean floor[1] and anoxic waters[2] fishes make a living throughout most of the deep-sea environment. Considered as a whole, the deep ocean

1. The sole-like "fish" sighted by Jacques Piccard from a bathyscaphe at about 10,000 meters in the Challenger Deep may have been some kind of holothurian (Wolff, 1961).
2. In the Cariaco Trench off Venezuela, Mead (1963) took numerous *Bregmaceros atlanticus* between depths of 400 and 600 meters, where there is virtually no dissolved oxygen.

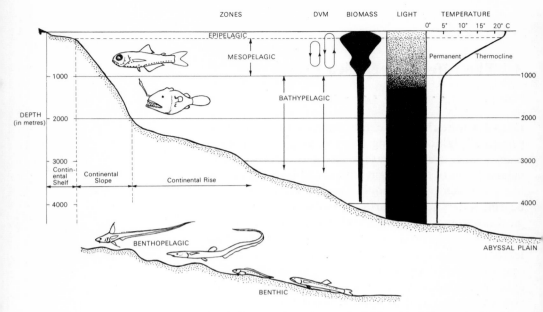

10   A diagram of certain oceanic features in relation to the life of deep-sea fishes: mesopelagic (represented by a lantern-fish); bathypelagic (by an anglerfish); benthopelagic (by a rat-tail, left, and by a halosaur, right; benthic (by a sea-snail, left, and *Bathymicrops*, right). At the right of the diagram are represented the extent of diurnal vertical migrations (*DVM*) in the mesopelagic zone, the biomass of zooplankton, the light regime, and the temperature profile of the warm ocean.

impresses the biologist by certain vertical gradients and contrasts in light intensity, temperature, pressure, rate of circulation, nutrient-salt, and oxygen contents. When conceived as a series of environments in depth, marked by certain physical features and animal communities, the overall impression is of increasing monotony in the conditions of life as one moves into the depths. Mesopelagic and slope-dwelling fishes live in the most variable deep-sea environments. Below a level of about 1,000 meters, physical and biological changes are much less pronounced. Indeed, a depth of 1,000 meters may be taken as the rough dividing line between mesopelagic and bathypelagic faunas, the first much more diverse than the second. Fishes fit this scheme so well that one may wonder whether the transition from one fauna to another is related to well-marked environmental changes, physical or biological, or both. The threshold of light, is an obvious possibility. But let us first consider the temperature structure of the ocean.

Deep-Sea Fishes

Under the warm waters between latitudes 40°N and 40°S, where most deep-sea fishes grow and reproduce, temperatures trace a characteristic profile. From the near-surface ("seasonal") thermocline, temperatures fall rapidly from values near 20°C to about 4 to 8°C at 1,000 meters (Figure 10). The level of most rapid fall is the axis of the main or permanent thermocline, which is deep enough to be beyond the reach of seasonal changes. This thermocline underlies the central water masses, and it is deepest under regions of greatest heating and evaporation. "In the North Atlantic the 10°C isotherm, which has been considered to be near the centre of the main thermocline layer, is found at a depth near 800 to 900 metres just south of Bermuda, and near the 500-metre level in the central South Atlantic. In the Pacific Ocean, the same isotherm is nowhere deeper than 500 metres" (Von Arx, 1962, p. 197). From 1,000 meters to the mean depth of the ocean (4,000 meters) there is a very gradual fall of 2 to 5°C. The slope of the main thermocline is so steep that one almost expects it to be a faunal barrier—at least, for certain species. Bruun (1957), indeed, regarded the 10°C isotherm as the dividing line between mesopelagic and bathypelagic faunas—between animals of the (warm) thermosphere and the (cool or cold) psychrosphere. There may well be different communities of mid-water fishes on either side of the permanent thermocline, but both are generally classed as mesopelagic species. For instance, above are centered the populations of many kinds of myctophids and stomiatoids (for example, *Vinciguerria* spp., *Maurolicus* spp., astronesthids, melanostomiatids, and so on); below, down to near the 1,000 meter level, are certain hatchet-fishes, pale-colored *Cyclothone* spp., malacosteids, and so forth. Deeper than 1,000 meters is a fauna dominated in species by angler-fishes (ceratioids) and in numbers of individuals by dark species of *Cyclothone*. It is these fishes and others that form a bathypelagic fauna, one with many unifying features.

In subtropical and tropical oceanic regions, a depth of some 1,000 meters seems to be the upper limit of the vertical range of most ceratioid angler-fishes (Bertelsen, 1951). This limit is hardly marked by the 10°C isotherm, which varies greatly in level over much of the ocean. For instance, between latitudes 40°N and 40°S and along 160°W in the Pacific Ocean, the depth of the 10°C isotherm ranges from about 100 to 500 meters (Reid, 1965). There is certainly a nearer fit between the threshold of light and the transition from mesopelagic to bathypelagic faunas of fishes. But how good is this fit?

In the clearest parts of the ocean, sunlight can be detected and measured by sensitive equipment down to depths of about 1,000 meters. Recently, Clarke and Kelly (1964) measured light penetration and transparency along transects in the western Indian Ocean. So clear was the water at one station (154) that the irradiance at a depth of 900 meters was 6×10^{-9} percent of the surface value, equivalent to an intensity of about 6×10^{-6} $\mu w/cm^2$ at noon. They continue: "If we take 3×10^{-16} $\mu w/cm^2$ as the threshold for vision for deep-sea fishes (Clarke and Denton, 1962) and if we extrapolate the curve for Station 154, we find that a deep-sea fish could probably detect the presence of daylight at 1,300 m at noon in this part of the Indian Ocean." At the least transparent position similar procedure gave them 700 meters for the limit of fish vision.

This order of variation hardly seems to make the threshold of light a stable level of reference. But data from the above paper and from one by Clarke and Backus (1964) show that waters below 100 meters are relatively and rather uniformly clear. All measurements made by Clarke and his colleagues "in any part of any ocean within the stratum from 300 to 1,000 m. have yielded consistently low attenuation coefficients which fall within the range K = .021 to .042" (Clarke and Kelly, 1964, p. 11). In oceanic waters, then, changes in transparency from place to place are largely in the surface layer—and are evidently due to changes in the content of fine particles. At any one place particle-screening is likely to vary from time to time, though "judging from data so far collected, no drastic changes appear in the average optical characteristics of particles from one oceanic area to another" (Jerlov, 1966, p. 96). Fluctuations in the threshold of visibility may thus center about a mean depth, which may well be close to 1,000 meters over much of the environment of deep-sea fishes.

On crossing this threshold, some change in eye structure might well be expected, but not necessarily in faunal structure. Actually, there are changes in both. First, though, we will consider the former in a general survey of sense organ design.

Deep-Sea Fishes

THE SENSORY LIFE OF DEEP-SEA FISHES

The eyes of mid-water fishes. Most kinds of mesopelagic
fishes have prominent to large eyes; a few, such as the
pale-coloured *Cyclothone*, have small eyes, but no species is known to have regressed eyes. The populations of
mesopelagic fishes are centered between 200 and 1,000
meters, and they belong to the following main groups
(see also Appendix):

Order Isospondyli (Figure 11)
   Suborder Stomiatoidea (Gonostomatidae, Sternop-
   tychidae, Astronesthidae, Chauliodontidae, Mela-
   nostomiatidae, Stomiatidae, Idiacanthidae,
   Malacosteidae)
   Suborder Clupeoidea (Searsidae, Alepocephalidae)
   (some species live near the bottom)
   Suborder Argentinoidea (Microstomatinae,
   Bathylagidae, Opisthoproctidae)

Order Iniomi (Figure 12)
   Suborder Alepisauroidea (Scopelarchidae, Evermannellidae, Omosudidae, Anotopteridae, Alepisauridae, Paralepididae)
   Suborder Myctophoidea (Myctophidae, Scopelosauridae)

Order Giganturoidea (Giganturidae; Figure 12)
Order Miripinnati (Mirapinnidae, Eutaeniophoridae)
Order Apodes (Nemichthyidae, Nessorhamphidae)
Order Anacanthini (Melanonidae)
Order Berycomorphi (Melamphaidae, Diretmidae,
Anoplogasteridae; Figure 13)
Order Allotriognathi (Stylephoridae)
Order Percomorphi (Gempylidae, Trichiuridae,
Chiasmodontidae; Figure 13)

With the notable exception of certain alepocephalids,
gonostomatids, and trichiurids, virtually every member
of the above group is a mesopelagic fish.

Concerning the many species with medium to large
eyes, some, such as certain lantern-fishes (myctophids)
and barracudinas (paralepidids), seem to have eyes much
like those of fishes from shallower waters. It is only on
closer scrutiny that one finds certain trends in the eye design of mesopelagic fishes. Even if the eye is not large, the
pupil is relatively wide and stopped by a large lens. Eyes
of this kind, compared to others of the same size with a
smaller pupil and lens, allow more light quanta to enter

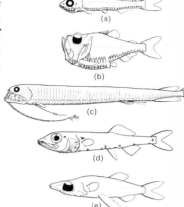

11   Some isospondylous fishes of the
mesopelagic zone. (a) *Vinciguerria*
(Gonostomatidae); (b) a hatchet-fish,
Argyropelecus affinis (Sternoptychidae) and (c) *Flagellostomias boureei*
(Melanostomiatidae), belong to the
suborder Stomiatoidea; (d) *Persparsia
tåningi* (Searsidae); (e) *Rhynchohylus*,
Opisthoproctidae.

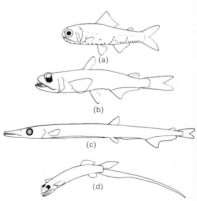

12   Mesopelagic fishes. (a) *Diaphus
effulgens,* a lantern-fish (Myctophidae);
(b) *Scopelarchus candelops* (Scopelarchidae); (c) *Lestidium atlanticum*
(Paralepididae); (d) *Gigantura chuni*
(Giganturidae). [(b) and (c) are
redrawn from Rofen, 1966.]

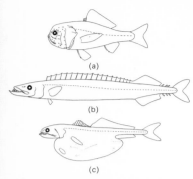

13 Spiny-finned mesopelagic fishes.
(a) *Melamphaes lugubris* (Melamphaidae); (b) *Thyrsitops violaceus*
(Gempylidae); (c) *Chiasmodon niger*
(Chiasmodontidae), giant-swallower.
[(a) is redrawn from Ebeling, 1962;
(b) and (c) are redrawn from Goode
and Bean 1896.]

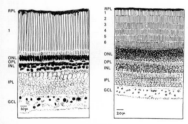

14 Sections through the retina of
Vinciguerria lucetia (left), a gonostomatid fish (see Fig. 11), and an
argentinoid fish, *Nansenia groenlandica* (right). The latter has six (1–6)
tiers of acromeres; the former, one.
RPL, pigment layer of retina; *ONL*,
outer nuclear layer; *OPL*, outer plexiform layer; *INL*, inner nuclear layer;
IPL, inner plexiform layer; *GCL*,
ganglion cell layer. (Redrawn from
Munk, 1966.)

the eye. The larger the absolute size of the eye and the
greater the relative size of its pupil and lens, the better it
is for gathering and registering the light of small bioluminescent sparks. In the twilight zone both sets of
adaptations are biologically desirable and each species
has presumably evolved some kind of compromise eye.
"But, whatever the structure and use of the eye the efficiency with which the retina absorbs the light quanta in-
cident on it will be of the first importance" (Denton and
Warren, 1957, p. 651).

Apart from *Omosudis*, which has a curious, cone-rich
retina (Munk, 1965), all deep-sea fishes have pure-rod
retinae. Compared to fishes from shallow waters, the
rods of mesopelagic fishes tend to have longer light-
absorbing parts (acromeres). Indeed, certain argentinoid
and stomiatoid fishes, for instance, *Nansenia groenlandica*, *Bathylagus* spp., *Winteria*, *Bathylychnops*, *Gonostoma elongatum*, *Chauliodus sloani*, *Stomias* spp., and
Eustomias obscurus, have two or more layers of rod
acromeres in the retina (Figure 14 and Munk, 1966).
Such elaboration, compared to that of mesopelagic species with a single-layered retina, does not necessarily
increase the length of the light-absorbing region. In
general, though, light reaching the retina of a mesopelagic fish moves into a thick close pile of light-absorbing
elements. Summation of visual response, indicated by the
mean number of rods linked to one optic nerve fiber,
seems quite variable, but present data are not adequate
(Munk, 1965, 1966).

Whatever the fine design, the eyes of mesopelagic
fishes, such as *Argyropelecus* spp., *Chauliodus sloani*,
Gonostoma elongatum, *Searsia* sp., *Myctophum puncta-
tum*, and *Diretmus*, have a high density of photosensitive
pigment compared to the human eye. The human retinal
density of rhodopsin for light of 500 mμ which is about
0.15, may be compared with the figure of 1.0 or more for
a number of mesopelagic fishes. If this visual adaptation,
together with the advantages the fish has in a wider pupil
and more transparent eye media, are considered together, "we may suppose that if a deep-sea fish and a
human were both looking at the same large field of
blue-green light, the number of quanta absorbed/cm^2 of
retina/sec. would be 15–30 times greater for a deep-sea
fish than for the human" (Denton and Warren, 1957).
On these and other estimations are based figures in-

Deep-Sea Fishes

dicating that deep-sea fishes can see daylight at depths below 1,000 m in the clearest parts of the ocean.

Such sensitive, wide-open eyes can be improved. The pupil may be enlarged so that it is not entirely stopped by the lens, thus leaving an aphakic (lensless) space. The pupillary rim of the iris may be more or less concentric with the outline of the lens, as in *Stomias* spp. and certain myctophids, or eccentric. In many of the eccentric forms the aphakic space is before the lens and the pupil is a circular to oval aperture (Figure 15). This kind of eye has evolved in searsids, bathylagids, malacosteids, scopelosaurids, certain myctophids (*Diaphus* spp., *Taaningichthys* spp., *Lampadena* spp.), and so forth. Munk (1966) has studied the eyes of *Nansenia groenlandica, Bathylagus pacificus, Platytroctegen mirus, Eustomias obscurus*, and free-living males of the angler-fishes *Cryptopsaras couesi* and *Ceratias holboelli*, which all have rostral aphakic spaces. As Munk says, these spaces expand the forward, binocular field of vision, which may also be enhanced by a sighting groove before each eye. Munk found these grooves in *Eustomias* and the two angler-fishes; they are also present in *Scopelosaurus* spp.

The binocular field covers a specially modified, temporal part of each retina. In *Nansenia, Bathylagus, Eustomias*, and the two angler-fishes, the rods of this temporal area have the longest light-sensitive parts but Munk does not say whether there is also an increase in visual-cell density. At all events, the binocularity and the (temporal) lengthening of the acromeres should give high sensitivity in the rostral field of vision, coupled with good perception of objects in aquatic space. Both faculties are biologically valuable in dim surroundings.

In searsids, bathylagids, and *Scopelesaurus* spp., the specialized, temporal area of the retina has contracted, as it were, to a fovea (see Marshall, 1966; Munk, 1966; Vilter, 1954). A line from the fovea through the center of the lens and the aphakic space marks the keen visual axis of the fish, which again has a wide binocular field. The striking part of the foveae of these deep-sea fishes is not the depression, which may be absent, but the very long visual segments of the rods. The fovea is, presumably, the most sensitive part of the retina. Though cone-rich foveae are known in diverse teleosts from shallow waters (Walls, 1942), it is still remarkable to find even a pure-rod fovea in a deep-sea fish. But at certain mesopelagic

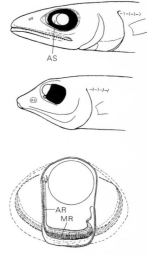

15   Eyes of mesopelagic fishes. Above: *Scopelosaurus*, with an egg-shaped eye and pupil and aphakic space *AS*. Middle: *Winteria telescopa*, with forwardly directed tubular eyes. Below: a tubular eye in relation to a normal eye. *AR*, accessory retina; *MR*, main retina.

levels above 1,000 m, bioluminescent flashes may at times be frequent enough to merge into a nearly continuous background of light. In such surroundings a fovea could even be used to scan organisms of interest to its owner.

Thus, in some mesopelagic fishes a certain extra-sensitive part of the retina, whether a temporal area or fovea, is visually correlated with a rostral aphakic space in the eye. But quite a number of mesopelagic fishes have a ventral aphakic space, which has not been investigated. Munk (1965) has drawn attention to such a space in *Omosudis lowei*, and it is also found in numerous lantern fishes (for example, *Electrona antarctica*, *Myctophum affine*, *Hygophum reinhardti*, *Tarletonbeania crenularis*, *Lampanyctus leucopsarus*, and *L. mexicanus*). Judging by the excellent illustrations in Rofen (1966), some of the barracudinas also have a ventral aphakic space (for example, *Paralepis atlantica*, *P. elongata*, *Notolepis rissoi*, and *Lestidiops mirabilis*). This suggests that the line of keenest vision is downward, though we should remember that such fishes may often be poised at angles away from the horizontal plane. In any event, study of the lens mechanism and fine retinal structure of these fishes ought to be undertaken.

Through quite a different trend of ocular evolution, diverse mesopelagic fishes have gained even greater binocular vision than have species with an aphakic complex. Tubular eyes, each fully stopped by a relatively large lens, and with the two main visual axes virtually parallel, have been independently acquired by members of eleven families. These are:

Isospondyli
  Stomiatoidea
      1. Gonostomatidae: *Ichthyococcus* only
      2. Sternoptychidae: *Argyropelecus* only (Figure 11)
  Argentinoidea
      3. Argentinidae (Microstomatinae): *Xenophthalmichthys* only
      4. Opisthoproctidae: all except *Bathylychnops* (Figures 11 and 15)
Iniomi
  Alepisauroidea
      5. Scopelarchidae: all species (Figure 12)
      6. Evermannellidae: all except *Odontostomops*

Myctophoidea
   7. Myctophidae: *Hierops* only
Giganturoidea
   8. Giganturidae: all species (Figure 12)
Apodes
   9. Family unknown: *Leptocephalus mirabilis* Brauer
Allotriognathi
   10. Stylephoridae: *Stylephorus*
Pediculati
   Ceratioidea
     11. Linophrynidae: metamorphosed, free-living
       males (Figure 16)

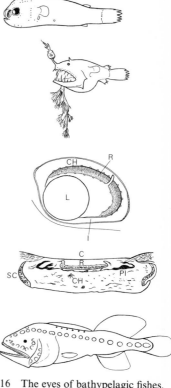

In tubular eyes the iris is much reduced[3] and at the end of the tube is a main retina, which continues over part of the walls, close to the lens, as an accessory retina. The main retina has much longer rods than the accessory extension. These features and others were first closely studied by Brauer (1908), but recently Munk (1965, 1966) has reviewed this work in the light of his own research.

The wide binocularity of tubular eyes should enable a fish to judge the distance of nearby organisms—an obvious advantage in dim surroundings—and may also provide extra sensitivity (as do human eyes when used together). But most of the sensitivity resides in the main retina, with its very efficient light-absorbing screen of golden pigments. (See Denton and Warren's (1957) figures for the density of pigment in the eyes of *Argyropelecus*). Further, the enlargement of an image on the main retina is the same as that in a normal eye of the same dioptric dimensions (Figure 15). Such a saving of ocular space—and most mesopelagic fishes are small—is gained at the expense of the accessory retina, which, as Munk (1965, 1966) demonstrates, is too close to the lens to be in focus. Tubular eyes are thus not bifocal. Even so, the accessory retina, as Brauer (1908) believed, could still detect movements, though only of brightly lit organisms.

The multiple convergent evolution of these extraordinary eyes would almost seem to proclaim some outstanding biological significance. But tubular eyes are not clearly correlated with form or a special mode of life. All short, deep-bodied forms, such as *Argyropelecus, Ichthyococcus, Opisthoproctus,* and *Macropinna*, certainly have upwardly turned eyes. Some with fusiform or elongated

16 The eyes of bathypelagic fishes. Above: male (left) and female (right) of the ceratioid anglerfish, *Linophyrne arborifera*. (Redrawn from Bertelsen, 1951.) Below the female is an oblique horizontal section through the tubular eye. *L*, lens; *CH*, choroid; *R*, retina; *I*, iris. (Redrawn from Munk, 1966.) The lowermost fish is a whale-fish, *Ditropichthys storeri*, above which is a section cut through one of its very regressed eyes. *C*, cornea; *R*, retina; *PI*, retinal pigment; *SC*, sclerotic cartilage. (Redrawn from Munk, 1966).

3. Not in male *Linophryne* (Munk, 1964).

bodies (*Winteria, Xenophthalmichthys, Gigantura, Bathyleptus*, and *Stylephorus*) have forwardly directed eyes; others of the same range of form (*Dolichopteryx*, Scopelarchidae, and Evermannellidae) have upwardly looking eyes (Figures 11 and 12). If tubular eyes are particularly good for seeing prey silhouetted against the relatively strong, down-going part of the light field—and silhouette disguise by ventral light organs may well be advantageous for mesopelagic organisms—then the fishes with forwardly directed eyes should spend some of their time hovering head upwards in the sea. Such a habit is known in myctophids and paralepidids, and has evidently been assumed for giganturids and *Stylephorus* (see Figure 1 in Bruun, 1957).

But there is no doubt that tubular eyes are not correlated with a special regimen of their owners. Some forms, such as *Opisthoproctus* and *Macropinna* have tiny, "nibbling" jaws; others, such as evermannellids and giganturids, are large-mouthed predators, able to swallow large prey. There is, however, a correlation that is becoming more evident: tubular-eyed fishes tend to live in the lower reaches of the mesopelagic (twilight) zone.

Curiously enough, Hjort (1912) stated that tubular eyes were found in fishes from depths less than 500 meters. But Jespersen and Tåning (1926) noticed that *Vinciguerria attenuata*, which has incipient tubular eyes, most pronounced in the adolescent phase, lives below a related species (*V. poweriae*) with normal eyes. The eyes of the first species seemed to fit its (dimmer) surroundings but close comparative study of both has yet to be undertaken. *Ichthyococcus*, a gonostomatid with fully developed tubular eyes, spends its adult life largely at levels near 500 meters. In the Mediterranean at least, this is also true for *Argyropelecus hemigymnus* (Jespersen, 1915). Recent surveys with closing nets in the Canary Basin reveal that *Argyropelecus* lives at levels of 500 m and below. "At 600–700 m many more tubular-eyed fishes appear. To the hatchet fishes may be added the argentinoid *Dolichopteryx* and pale coloured species such as *Scopelarchus* and *Evermannella*" (Harrisson, 1967, p. 80). Species with a well-marked rostral aphakic space, for instance, malacosteids, lantern-fishes of the genera *Lampadena* and *Taaningichthys*, certain bathylagids, and searsiids, also tend to live at lower mesopelagic levels.

All this suggests that tubular eyes and eyes with an

aphakic complex were independently evolved to enable their owners to live in the dimmer parts of the twilight zone. Further, it seems significant that tubular-eyed fishes, if they undertake diurnal migrations, do not appear in the surface layers at night (Marshall, 1960). Since the components of deep scattering layers—and *Argyropelecus* spp. are probably involved—keep to certain isolumes, these fishes could be tied to the dimmer levels. But are tubular eyes more sensitive than dioptrically equivalent normal eyes? Judging from the figures for *Argyropelecus* spp., tubular-eyed fishes tend to have higher densities of photosensitive pigments than normal-eyed, mesopelagic fishes (see table in Denton and Warren, 1957). There may also be the extra sensitivity given by binocular vision. Thus, tubular-eyed fishes probably have very sensitive vision, coupled with means of judging distances. Both attainments are obviously valuable to predators, but what of *Opisthoproctus*, which is known to browse on siphonophores (Marshall, 1954). *Opisthoproctus* should be able to visually mark the position of luminescing siphonophores, which are by no means quiescent organisms. Bertelsen and Munk (1964) believe that the eyes of an *Opisthoproctus* detect—and presumably pinpoint—recognition signs emitted by the ventral light organ system of another individual of the same kind.

The eyes of bathypelagic fishes. Tubular eyes are not confined to mesopelagic fishes: they also develop on free-living, metamorphosed males of the ceratioid anglerfish family, Linophrynidae (Figure 16). Like the females, the males of ceratioids are commonest at depths of 2,000 meters or more (Bertelsen, 1951), where perpetual darkness is broken by sparse flashes of bioluminescence. Free-living males of other families (Melanocetidae, Himantolophidae, Oneirodidae, and Ceratiidae) have a rostral aphakic complex in the eyes, most strongly developed in the ceratiids (Bertelsen, 1951). Munk (1964, 1966) has studied the eyes of male *Linophryne arborifera*, *Cryptopsaras couesi*, and *Ceratias holboelli*, the last two being ceratiids. The forward binocularity of all three species is visually matched by a special, temporal part of the retina, which bears the longest rods. Such sensitive, distance-judging eyes should not only help their owners to catch luminescing prey but also lead them finally to the flashing lure of their female partners. Male angler-fishes with minute, even degenerate, eyes (for example, gigan-

17 Bathypelagic fishes. (a) *Cyclothone*; (b) *Saccopharynx harrisoni,* a gulper-eel; and two female ceratioid anglerfishes, (c) *Himantolophus groenlandicus* and (d) *Danaphryne nigrifilis.* [(c) and (d) are redrawn from Bertelsen, 1951.) Note the small eyes.

tactids) may well rely most on their large olfactory organs to find their mates.

The eyes of female ceratioids grow very little after metamorphosis (Bertelsen, 1951). Thus, their eyes are quite small (Figure 17), though Munk (1964) found no signs of degeneration in fairly large ceratiids and *Galatheathauma*. But at best, these eyes can do no more than record adjacent luminescence. (In the ceratiids the lens is even behind the iris.)

Both in numbers and the extent of their living space, the dark species of *Cyclothone* are dominant, bathypelagic fishes (Figure 17). Except for *C. obscura,* which has minute eyes, fully grown individuals have eyes from 0.5 to 1.0 mm in diameter. Judging from Brauer's (1908) report, the eyes are in no way regressed. The wide open pupil and large lens, which fills most of the posterior chamber, remind one of the eye of a mouse or a shrew. In fact, the smallest shrews have eyes of about the same dimensions as the upper limit of eye size in *Cyclothone*.

Though they have a wide pupil, the eyes of *Cyclothone* are too small to do more than gather nearby luminescence. Can one *Cyclothone* peer closely at another for "recognition" lights? Each species may well have more luminescent individuality than we can see on dead specimens. And *Cyclothone obscura,* with the minutest eyes, has virtually lost its photophores.

Among members of the bathypelagic fauna are also *Gonostoma bathyphilum,* the gulper eels (Lyomeri; Figure 17), the eels *Cyema* and *Avocettina,* the aphyonine brotulids, certain macrourids (*Odontomacrurus, Cynomacrurus, Echinomacrurus,* and *Squalogadus*), and probably, most of the whale-fishes (Cetunculi). Of these, the eels and gulper eels have "bull's-eye" eyes, seemingly like those of *Cyclothone*. *Gonostoma bathyphilum* has small eyes with a wide open pupil and small lens (Marshall, 1954). Apart from the dioptric aberration, these eyes appear well formed, but Munk (1966) found a regressed retina and a lens defect in *Echinomacrurus mollis,* again with seemingly good eyes. Evidently, external form can be misleading. Much greater regression occurs in the eyes of some whale-fishes (for example, *Ditropichthys* and *Gyrinomimus*) and in all aphyonine brotulids (Brauer, 1908; Munk, 1966; Figure 16).[4]

4. The aphyonine brotulids are *Aphyonus, Sciadonus,* and *Barathronus*.

46

Deep-Sea Fishes

We shall see later that there are certain parallels between the development of eyes in mid-water and bottom-dwelling, deep-sea fishes. First, though, and by way of summary, the two mid-water faunas will be compared and contrasted. The eyes of mesopelagic fishes are designed to see in dim or very dim light and to register luminescent displays, which may be very frequent. No doubt some of these displays are recognition signs between members of the same species. (One thinks particularly of the lantern-fishes.) But there is no relation between ocular elaboration and the provision of light organs. There are both luminescent and nonluminescent fishes with large, normal eyes, tubular eyes, and eyes with a rostral aphakic complex. Consider, though, that the entire mesopelagic fauna is involved in some way in diurnal vertical migrations. To detect diurnal changes of illumination in the twilight zone calls for large, wide open, and sensitive eyes. The eyes of migrating species must signal the onset of sunset and sunrise, the times when they presumably "follow their isolume" upwards or downwards.

The bathypelagic environment is sunless, but there is sporadic luminescence. Some flashes come from the lures of female angler-fishes, and we saw that the males of most species have the right kinds of eyes to find the lights of their partners. Except for these males, eyes are relatively small or regressed in members of the bathypelagic fauna. But small complete eyes, like those of most *Cyclothone*, may detect simple luminescent events. These and earlier considerations suggest that in sunless waters eyes are kept if there is "meaningful" luminescence to be seen. But the most elaborate eyes—those of dwarf male ceratioids—are quite small. Evidently, there are not enough "stars" to make large "telescopic" kinds of eye worth while. After all, when eyes regress, orientation and food-finding in fishes can be centered in the acoustico-lateralis and olfactory systems. Crustaceans, most of which have olfactory hairs and a distance touch sense in the form of antennae, and so on, are also able to live without eyes in the sunless deeps. But cephalopods, with no kind of lateral line system and virtually no olfactory organs[5] retain well-developed eyes in all parts of the deep sea.[6]

5. Except in *Nautilus*.

6. *Cirrothauma* and *Cirroteuthis* are the only cephalopods known to have regressed eyes.

The eyes of bottom-dwelling species. Hundreds of photographs and numerous observations from underwater vehicles have shown that some fishes swim or hover near the deep-sea floor. Most of these *benthopelagic* species have a swimbladder (Marshall, 1960), and the most diverse groups are rat-tails (Macrouridae), deep-sea cods (Moridae), and brotulids. Halosaurs and notacanths (Heteromi), again with a swimbladder, also move over the oozes. The other bottom-dwelling fishes, which have no swimbladder, have the habit of resting on the deep-sea floor. The most diverse of these *benthic* forms are sea snails (Liparidae), eel pouts (Zoarcidae), Chlorophthalmidae, and tripod fishes (Bathypteroidae). The curious ipnopids also belong here.

Survey of the rat-tails and brotulids (Marshall, 1954, 1965a) or of large genera, such as *Careproctus* (Liparidae) and *Lycenchelys* (Zoarcidae) (Rass, 1966), shows that species with populations centered between depths of 200 and 1,000 meters have large eyes, whereas those that live below 1,000 meters have small or regressed eyes (Figures 18 and 19). Considering first the deeper dwellers, regressed eyes are known in certain brotulids (for example, *Bassogigas profundissimus, Leucicorus lusciosus, Typhlonus*), bathypteroids (*Bathypterois longipes, Benthosaurus grallator*), and liparids (*Careproctus kermadecensis*). (See Thinés, 1955; Munk, 1966.) In these species the lens is faulty, reduced, or absent and the retina shows varying signs of degeneration. The eyes of the deeper-living macrourids (for example, *Chalinura, Lionurus,* and *Nematonurus*) have yet to be examined, but arguing from the bathypelagic *Echinomacrurus* one might expect some regression in the eyes of its benthopelagic relatives from the sunless depths. One striking feature of a moving film taken from a bathyscaphe at depths around 2,500 meters is that macrourids and a halosaur seem to be quite unaffected by the lights of the vehicle.

In the deep sea most kinds of bottom-dwelling fishes live over the slope at depths less than 1,000 meters. The eyes of such fishes have not been well studied, but Brauer (1908) made a beginning with certain macrourids, a brotulid, and an eel (*Coloconger*). Yet it is clear that the eyes of slope-dwelling fishes parallel those of mesopelagic species in their relatively large dioptric dimensions and pure-rod retinae. But more specialized eyes are rare. No species has tubular eyes, and eyes with a rostral aphakic

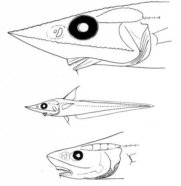

18   Eye size in a slope-dwelling rat-tail. *Coelorhynchus occa* (the two upper figures) and an abyssal rat-tail, *Lionurus filicauda* (bottom).

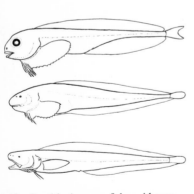

19   Benthic deep-sea fishes. Above: sea-snail, *Careproctus cypselurus* (Bering Sea, 930–1620 meters, with relatively large eyes; middle: *Careproctus amblystomopsis,* taken at a depth of 7230 meters in the Kurile-Kamchatka Trench, with minute eyes; bottom: a brotulid, *Bassogigas profundissimus* (7130 meters), also with very small eyes.

complex are known only in chlorophthalmids.[7] In fact, species that live in the lower reaches of the twilight zone develop smaller eyes than those from the upper reaches. In the Philippines region, for instance, species of the macrourid genera *Gadomus* and *Bathygadus*, which occur mostly at depths below 600 meters, have eyes less than 18 mm in diameter. In species of *Coelorhynchus* from depths less than 600 meters the eyes are 20 to 30 mm in fully grown individuals (Gilbert and Hubbs, 1920). In the sunless depths beyond 1,000 meters there are no known exceptions to the rule of regressed eyes. But a bottom-dwelling fish can even more readily dispense with sight than a mid-water form. Besides its olfactory and acoustico-lateralis organs, it can use its gustatory system. For instance, macrourids, brotulids, and morids should have an extensive outer coverage of taste buds, judging from the course of the relevant (ramus lateralis accessorius) nerves (Freihofer, 1963). Moving benthic organisms may also be found by tactile means. "Life on the bottom is largely life in one plane, and the finding of food by touch and chemoreception is vastly easier. Go far enough along the bottom (if you're a fish), and you're bound to bump into something good to eat" (Walls, 1942, p. 398).

Olfactory organs. Sensitive eyes with "localizing devices" have a very limited range, especially in dim surroundings—and if their structure has to be fitted on a small fish. Dwarf male angler-fishes with such eyes are mostly from 20 to 40 mm in length. A male will eventually close on the light of his mate at very close range: at even closer range he will presumably recognize the "design" of the light, which is peculiar to each species. Some male ceratioids have minute eyes, but the males of most species have large olfactory organs, ceratiids being among the few exceptions (Bertelsen, 1951).

After metamorphosis, the males rapidly become macrosmatic. When fully developed each olfactory organ contains a stack of broad sensory lamellae, which are open to the sea through two wide nostrils. The olfactory organs of female ceratioids are reduced to minute papillae. Matching this extreme dimorphism, the olfactory nerves, bulbs, and forebrain are strongly developed

---

7. North Atlantic species of *Chlorophthalmus* and *Parasudis* live between depths of 180 and 730 meters (Mead, 1966).

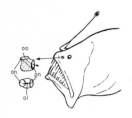

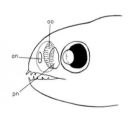

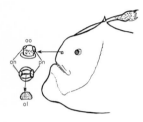

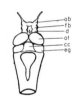

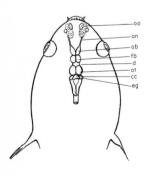

in males but regressed in females (Figure 20; Marshall, 1967).

Other bathypelagic fishes have "copied" these features of ceratioids. Males of *Cyclothone* species develop a much larger olfactory complex (that is, organ → nerve → olfactory bulb → forebrain) than the females. In fact, female *Cyclothone* have a very reduced olfactory system (Figure 21). Moreover, the male fish are smaller, but the difference is not so marked as in ceratioids. Mature males are mostly from one half to three fifths of the span of females, which when fully grown range from about 25 to 60 mm in the various species.

In *Gonostoma bathyphilum*, which is related to *Cyclothone* and has aberrant eyes, the males once more have much the larger olfactory organs. Again, the males are smaller—by a factor of half to three fifths—than the females, the length of which ranges to 200 mm or more.

Concerning the other bathypelagic fishes, female gulper-eels, bob-tailed snipe-eels (*Cyema*), and blind brotulids (*Aphyonus*) are certainly microsmatic. Nothing is known of the males, but both sexes of the eels *Nemichthys* and *Avocettina* have small olfactory organs (Marshall, 1967).

Ceratioid angler-fishes and *Cyclothone* spp., the two most successful groups of bathypelagic fishes, have thus evolved distinct sexual dimorphism in olfactory organization and size. But these differences are *not* found in *Gonostoma denudatum*, *G. atlanticum*, and *G. elongatum*, mesopelagic relatives of *G. bathyphilum*. In fact, the only mesopelagic fishes with *both* dimorphic features[8] are the pale-colored species of *Cyclothone* (e.g. *signata* and *braueri*). Of the bottom-dwellers, only the maturing males of halosaurs are known to develop larger olfactory organs than the females (McDowell, M.S.).

8. Dwarf males are known in *Idiacanthus fasciola*, but olfactory development is small in both sexes.

20 Olfactory organs of ceratioid fishes. From top to bottom: *Melanocetus murrayi*, female; *Himantolophus groenlandicus*, female; brain of *Melanocetus murrayi*, female; olfactory organ and tubular eye of *Linophryne macrorhinus*, free-living male; brain and olfactory organs of *Oneirodes* sp., free-living male. *an*, anterior nostril; *cc*, corpus cerebelli; *d*, diencephalon; *eg*, eminentia granularis; *fb*, forebrain; *ob*, olfactory bulb; *oe*, olfactory epithelium; *ol*, olfactory lamellae; *on*, olfactory nerve; *oo*, olfactory organs; *ot*, optic tectum; *pn*, posterior nostril. (From N. B. Marshall, 1967. Courtesy Zoological Society of London.)

Deep-Sea Fishes

Bathypelagic fishes are spread in the largest living space on earth. If, as seems most reasonable, female ceratioids and *Cyclothone* secrete pheromones[9] to attract their mates, why should so vast an environment have "favored" this kind of chemical signalling? During any time in the breeding season of a species, the changes of sexual encounter will depend on the population densities of ripe females and responsive males. Now, through far-reaching economies in their organization, bathypelagic fishes, though living in a food-poor environment, are quite numerous. Between depths of 1,000 and 2,000 meters, where they are most abundant, Bertelsen (1951) estimated that there may be no more than a mean distance of 30 meters separating one ceratioid from the next. Separation between members of a particular species will be more than this, but the figure will be least where this species is dominant. In its area of dominance any species of *Cyclothone* is likely to be at least ten times as abundant as the local ceratioid population, which suggests that individuals have a mean separation of 3 meters or less. Moreover, the distribution of both ceratioids and *Cyclothone* is likely to be clumped rather than random.

But what is the ratio between ripe males and females in a species' area or areas of dominance? In a local ceratioid population, females may well be represented by at least two year groups and males by one only. Thus, females are likely to outnumber the males. But the males reach sexual maturity soon after metamorphosis, whereas the females take much longer to mature. "Of most species not a single mature female has been caught, and even if a contributory cause may be that these are fished less effectively than the younger specimens, we can hardly suppose that the unoccupied mature females amount to more than 2–5% of the total stock of metamorphosed females." After reasoning thus, Bertelsen (1951) estimated that there would be at least 15–30 ripe or near ripe males to every ready female. Perhaps female *Cyclothone* also live longer than the males, for the proportion of the latter in catches is usually below a third of the total count. Again, ripe males easily outnumber the mature females by a factor of ten or more (Marshall, 1967).

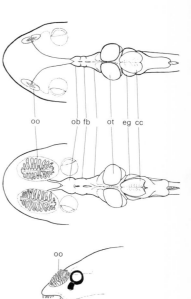

21   Olfactory organs and brain of *Cyclothone microdon*. From top to bottom; female; male; head of male. *cc*, corpus cerebelli; *eg*, eminentia granularis; *d*, diencephalon; *fb*, forebrain; *ob*, olfactory bulb; *oo*, olfactory organ; *ot*, optic tectum. (From N. B. Marshall, 1967. Courtesy Zoological Society of London.)

9. Pheromones are "substances which are secreted to the outside by an individual and received by an individual of the same species, in which they release a specific reaction, for example, a definite behaviour" (Butler, 1967).

Even if their olfactory thresholds are as low as that of freshwater eels, how might male ceratioids and *Cyclothone* find their mates? Much depends on the behavior of the females and on local water movements. At depths of about 2,000 meters in the North Atlantic current velocities vary from nearly zero to 15 cm/sec. with a mean of about 5 cm/sec. (Swallow, 1962). Dispersal of scents will thus tend to be rather slow.

In the sea, as in air, still conditions will confuse the keenest noses. But, given a breeze, male insects find a source of pheromones by flying upwind. Once a male has picked up the scent he turns into the wind, and as he flies "his olfactory receptors continue to be very slightly stimulated, keeping him flying upwind, until rather suddenly, when he gets within a few centimetres of the female, the stimulation greatly increases and indicates her proximity" (Butler, 1967).

But how might bathypelagic fishes, which cannot see or touch ground, face or swim into a current? Blind fishes show positive rheotaxis without tactile clues (Marshall and Thinés, 1958; Poulson, 1963). If we think of the mean current velocity (about 5 cm/sec.) at 2,000 meters in the North Atlantic, Poulson's observations on blind amblyopsid fishes are intriguing. He found that both species of *Amblyopsis* "often swim in midwater and show positive rheotaxis in currents of 2–7 cm/sec. without tactile reference". The biological value of such response in ceratioids and *Cyclothone* would seem to be great. If the semicircular canals of the ear are involved, they are certainly well developed in these fishes. Clearly, tests are needed, and female ceratioids, at least, may be caught alive.

Finally, there is the time factor. The food reserves of male ceratioids are mostly in the liver, which reaches maximum size at metamorphosis. Beside a sizable liver, *Cyclothone* males have substantial reserves of fat. While they are seeking their mates, the males can draw on these reserves for certain periods of time.

Thus, by visual and olfactory clues, males of the above groups may find their partners. It is certainly striking that olfactory dimorphism has evolved in luminous bathypelagic fishes. There is no evidence yet of such development in nonluminous species (Marshall, 1967). In the sunless bathypelagic region, pheromones, which can be perceived at greater range than visual signals, seem apt, considering the "favorable" factors of this environment.

Evidently, in mesopelagic surroundings light signals play the leading part in sexual attraction. For instance, the males of diverse stomiatoids (melanostomiatids, malacosteids, *Idiacanthus*) have larger cheek lights than the females. Luminous sexual differences in many myctophids involves the light organs on the caudal peduncle, and sometimes those on the head. But in slope-dwelling benthopelagic fishes, the trend, as we shall see, is decidedly towards a sonic sexual dimorphism. It must be remembered that the females of all species of bathypelagic fishes (and the males of a few) have regressed olfactory organs. These fishes cannot find their food by smell, and their eyes are very small or reduced. But hunting for food by scent is an energy and time consuming process and a hazardous way of making a living, particularly in a food-poor and sunless environment. Above all, fishes can "dispense with" olfaction because of their more reliable lateralis system—a means of finding prey that is moving nearby.

The lateral-line system. The sensory units of the lateralis system (neuromasts) are either housed in mucous-filled canals or freely exposed to the environment.[10] Whatever their situation, neuromasts have the same basic structure and function: they consist of hair-bearing sensory cells capped by a gelatinous cupula, the whole forming a receptor for displacements in the medium (Figure 22). In the words of Dijkgraaf (1962), lateral-line organs "serve mainly to detect and locate moving animals (prey, enemies, social partners) at short range on the basis of current-like water disturbances . . . Furthermore, there is evidence that the size of the moving object as well as its velocity and direction of movement are distinguished."

Most kinds of bathypelagic fishes, that is, ceratioid anglers, gulper eels (Lyomeri), and bob-tailed snipe eels (*Cyema*), just have free ending organs. Each neuromast is set at the end of an outgrowth of the skin that may be a low papilla, a short stalk, a flattened tag, or a long filament (Figure 23). Bathypelagic macrourids have free-ending organs and partly reduced canals. The wide canals of cetomimid whale-fishes, which house large neuromasts, are widely open to the sea through very large

22 The lateral-line system. Top: lateral-line canals on the head of a mesopelagic fish. *Searsia.* Middle: diagram of free-ending (*fo*) and canal organs (*co*), cupulae stippled. Bottom left: the free-ending organ of a larval teleost (After Iwai, 1967); *cu*, cupula; *sc*, sensory cells. Bottom right: organ from the head canal of a burbot, *Lota lota*; *n*, nerve. (Redrawn from Flock, 1965.)

10. A third group of ampullary lateral-line organs are electroreceptors (Dijkgraaf, 1962).

openings. Such canals are well on the way to obliteration. *Cyclothone* spp. certainly have large lateralis centers in the cerebellum, but no one has yet studied their lateral-line system.

Neuromasts set at the tips of long projections seem to be confined to bathypelagic fishes. Free-ending organs on cyclostomes, sharks, larval teleosts, and various adult teleosts are either slightly protuberant or flush with the surface of the skin. *Lophius* and the cave-dwelling amblyopsids, with neuromasts on low papillae, are among the exceptions (Figure 23).

Neuromasts in mucous-filled canals are buffered to some extent against "uninteresting" water displacements. Thus, they are largely shielded from displacements along the body surface, but they are well placed to be stimulated by local pressure differences, that is, by water displacements at right angles to the body surface (Dijkgraaf 1962). The cupulae of free neuromasts will be much more subject to water "noise," particularly to turbulent movements and to water flowing over the skin during active locomotion.

Fishes whose neuromasts are entirely or mainly free (for example, *Lepidosiren*, larval teleosts, cave-dwelling amblyopsids, gobies, *Lophius*, and so on) are slow or intermittent swimmers. Moreover, the first and third forms live in obscured or dark surroundings. For our present purpose, comparison of bathypelagic species with the amblyopsids is particularly relevant. In the true cave-dwellers, *Typhlichthys* and *Amblyopsis*, the neuromasts are extended on short stalks, but in the epigean species *Chologaster cornutus* and the troglophilic *C. agassizii* there is no such provision (Figure 23). Moreover, the cave dwellers carry more neuromasts and larger lateralis centers in the cerebellum than do *Chologaster* spp. When 40 mm *Typhlichthys* and *Amblyopsis* were stationary, their cupulae moved as adult water fleas (*Daphnia magna*) swam at distances of 15 to 35 mm from these receptors. Poulson (1963) concluded that "the 10–12 mm difference between the distance that a prey causes cupular movement in a stationary fish and the distance at which a gliding fish orients towards the same prey shows the advantage of decreased 'noise'."

Bathypelagic fishes live in waters that tend to be gentle in movement. The neuromasts of an angling ceratioid or gulper eel should thus be free to work at low levels of

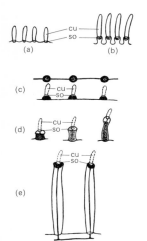

23 Free-ending lateral-line organs of two cave fishes [(a) and (b) redrawn from Poulson, 1963] and three ceratioid anglerfishes. (c) *Dolopichthys*, neuromasts on papillae; (d) *Cryptopsaras*, neuromasts on short stalks; (e) *Neoceratias*, neuromasts on long stalks, *cu*, cupula; *s*, sense organ (in *c* to *e* the cupula (broken line) has been assumed). Note that the true cave-dwelling fish *Amblyopsis spelaea* (b) has stalked sense organs, whereas those of its troglobitic relative, *Chologaster agassizi* (a) are almost flush with the skin.

water noise, a factor that also seems to "favor" olfactory signals between the sexes. Relevant experiments with an angler-fish, at least, are feasible. Meanwhile, we must note that the lateralis centers in the cerebellum and the lateral-line nerves are very well developed in ceratioids and other bathypelagic fishes (Marshall, 1967). Structurally, then, these fishes seem well equipped to detect and locate moving prey. Ceratioids feed on organisms ranging in size from copepods to prawns, squid, and fishes (Bertelsen, 1951). Bertelsen also observes that *Caulophryne* and *Neoceratias*,[11] the first with a non-luminous esca, the second without an illicium, have especially well-developed lateral-line organs. In both genera the neuromasts, which are carried at the end of long filaments, are thicker on the epidermal ground than they are in species with a luminous bait.

Of all deep-sea fishes, the bathypelagic assemblage is unique in this marked trend toward an exclusive system of free-ending lateralis organs. No mesopelagic or bottom-dwelling species is so made; these fishes have canal organs, though some also develop free-ending neuromasts. A mixed system is found in most bentho-pelagic fishes, but the main trend is toward wide cephalic canals containing large neuromasts, such as are found on macrourids, halosaurs, and many brotulids (Figure 24). The large, and presumably, very sensitive, organs of these fishes are well buffered against "extraneous noise" by a relatively large mass of viscous fluid. Moreover, judged by their form, and by photographs and ciné films, macrourids, halosaurs, and brotulids move slowly over the deep-sea floor in search of food.[12] Benthic deep-sea fishes, for instance, bathypteroids and ipnopids, have smaller canals and neuromasts than macrourids, but both groups are well provided with free-ending organs. Of the mesopelagic assemblage, relatively large neuromasts in wide canals are found on the heads of many myctophids and melamphiids (Marshall, 1954). But most kinds of deep-sea fishes live over the upper reaches of the continental slope or at mesopelagic levels, when there is likely to be more "water noise" than in the bathypelagic environment. We must also remember that many

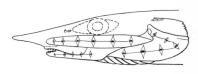

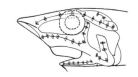

24 Wide head canals and large neuromasts in two benthopelagic fishes. Above: a halosaur, *Aldrovandia*. Below: a macrourid, *Coelorhynchus*.

11. The two long "nasal papillae" on the female appear to be much enlarged lateral-line organs (see Figure 30).

12. Other fishes with wide, macrourid-like canals, for example notopterids, are also slow movers.

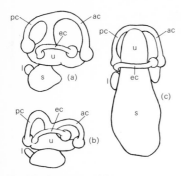

25 The inner ear of (a) *Vinciguerria lucetia,* a mesopelagic fish; (b) *Dolopichthys,* a ceratioid anglerfish from the bathypelagic zone; and (c) a macrourid, *Hymenocephalus.* (Redrawn from Bierbaum, 1914.) Note the enormous sacculus (*s*) of the macrourid. *u*, utriculus; *l*, lagena; *ac, pc, ec,* anterior, posterior, and external semicircular canals, each of which ends in an ampulla.

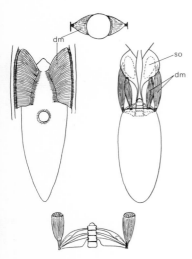

26 Sound-producing mechanisms on the swimbladder of a macrourid, *Malacocephalus laevis* (left), and a brotulid, *Monomitopus metriostoma* (right). The drumming muscles (*dm*) of the macrourid are inserted on the body wall and the forepart of the swimbladder; those of the brotulid attach to the swimbladder and modified ribs (shown below) and the otic capsules. Note the large saccular otoliths (*so*).

mesopelagic fishes hunt by night for their food in turbulent surface waters.

The inner ears. Fishes use down-welling light to keep a visual watch on their position of equilibrium. They also determine their orientation through the utricular chambers of the ears, which act as static organs. Any movement away from an even position produces a shear between the utricular otolith and its sensory macula. More precisely, the shearing component of gravity is the receptor-adequate stimulus to subsequent adjustments of position (Pfeiffer, 1964).

Both ways of keeping equilibrium are open to mesopelagic fishes. With regard to their visual means, they must depend on the downward component of the light field, which is registered by their sensitive eyes. Fishes of the sunless bathypelagic regions have only their static receptors for maintaining a position of equilibrium, and the same is also true of various small-eyed noctural species that live on the bottom in shallow waters.

Bathypelagic fishes certainly have well-developed utriculi and semicircular canals. This is particularly noticeable in small fishes, such as ceratioid males and individuals of *Cyclothone* species, in which well over half the volume of the neurocranium is devoted to the otic capsules. The otic capsules are largely taken up by the canals and the utriculus. Like the cave-dwelling amblyopsids (Poulson, 1963), these deep-sea fishes must depend greatly on the pars superior of the ear for monitoring their changes of position and angular accelerations.

Mesopelagic and bathypelagic fishes usually have a small sacculus (Bierbaum, 1914). But in most benthopelagic fishes, that is, macrourids, brotulids, and morids, this part of the ear is very large and so is the otolith (Figures 25 and 26). Since hearing resides in the saccular complex, it is surely significant that fishes of the above three groups have evolved sonic devices.

SOUND-PRODUCING MECHANISMS

Teleost fishes make sounds in two main ways: the action of relevant muscles either vibrates the swimbladder wall or causes stridulation between suitably shaped parts of the skeleton. Elasmobranchs, which have no swimbladder and an unsuitable skeleton, make no more than incidental sounds, as when they swim or feed (Marshall, 1962). But the gas-filled swimbladder found in many mesopelagic fishes is simply a hydrostatic organ; it is not provided with drumming muscles. Nor does it seem likely that these fishes have skeletal parts that are suitable for stridulation. This is true also of bathypelagic and benthic species, which in any event have no swimbladder (Marshall, 1960).

Except for the sharks, chimaeroids, and alepocephalids, benthopelagic deep-sea fishes have a well-developed swimbladder (Marshall, 1960, 1965a), but in the halosaurs and notacanths the swimbladder has no drumming muscles. This is also true of the deep-sea cods (Moridae), though they may well grate together their pharyngeal teeth, which are moved by large muscles that originate close before the forward wall of the swimbladder. Such stridulatory sounds, as certainly occur in certain haemulids and carangids, will be picked up and "given body" by the swimbladder. Moreover, the swimbladder forks anteriorly so that a circular pad on the forward part of each fork fits against a membrane-covered foramen in the rear wall of the otic capsule. The sacculus, as we have seen, contains a large otolith. Teleosts with some kind of coupling between the inner ears and the swimbladder have, at the very least, extra-sensitive hearing. But what do morids hear?

Most kinds of macrourids, like virtually all their relatives, the morids, are centered over the continental slope at depths between 200 and 1,000 meters. These slope-dwelling species, which also have a very large otolith in each sacculus, have evolved paired drumming muscles on the forward part of the swimbladder, but in male fishes only. Both attachments of each drumming muscle are either on the swimbladder wall, or one is on the adjacent body wall (Figure 26).

Brotulid fishes, also most diverse over the upper continental slope have more elaborate sonic devices. In the oviparous forms, which comprise about a hundred of some 175 species in all (Mead, Bertelsen, and Cohen, 1964), the male alone has large drumming muscles. These are attached to the otic capsule and the thick forward wall of the swimbladder. The compliance of the system probably resides in the modified ribs of the first three vertebrae, which converge to make forward suspensory bars for the swimbladder and to serve as points of attachment for all or part of the sonic muscles (Figure 26). The viviparous forms are less well known, but in some species, at least, both sexes have drumming muscles, and one or more pairs of the first three sets of ribs form attachments for the swimbladder and its muscles.

Since macrourids and brotulids are easily the most diverse groups of benthopelagic fishes in the deep sea, sound signalling is thus prevalent in the fauna of the upper slope. But in bathygadine group of macrourids, most of which tend to live at levels of 1,000 meters or more, and in the abyssal genera of the very diverse subfamily Macrourinae (for example, *Lionurus* and *Nematonurus*) neither sex has drumming muscles. Moreover, compared to their sonic relatives from the upper slope, these deeper-living macrourids have very small saccular otoliths. Sound making and large (hearing) otoliths would thus seem to be associated.[13]

More precisely, even without the deep-sea cods, well over half the benthopelagic fauna of the continental slopes (that is, some 300 macrourids and 150 brotulids) must be sound producers. In the macrourids and oviparous brotulids the male is the vocal sex, but note also that *both* sexes have large sacculi. In fact, sound making and correlated hearing is confined to species with large eyes, fitted for vision in a twilight world. Moreover, some of the morids and macrourids have a light gland on the belly. Most likely, sound signals and luminescence, if present, are involved in courting and mating activities. Male rivalry, as in the cod, a relative of the macrourids, may also be expressed in sounds. But why should abyssal macrourids, which have small, probably regressed eyes and no light glands, be without sound-making devices? One might think that in a sunless environment sound signals would be a good means of assembling the sexes. It seems, though, that directional help from a sound field is virtually confined to a (near-field) displacement region described by a radius of $\lambda/2$ from the source. Abyssal macrourids, as shown by a detailed photographic survey at depths around 2,500 meters, live in small groups (2 to 5 fishes), separated by much more than the radius of the near field.[14] Individuals of slope-dwelling species, which

13. This is also true in sciaenids (Schneider, 1962).

14. Investigations show that the fundamental frequency of swimbladder sounds correspond to the frequency of muscular contractions, which are not likely to be more than 150/sec.

can be very abundant, are probably very often within near-field earshot of one another. Even so, this gives us no inkling of how sexual encounter is managed in abyssal macrourids. And so far I have yet to find ovotestes in these forms, such as Mead (1960) and Nielsen (1966) have described for the bathypteroids and ipnopids.

But we can proceed a little and draw together the last three sections. Sound-producing fishes of freshwater and marine habitats tend to live near accessible interfaces between their medium and solid structures, such as the bottom, corals, rocks, weeds, and so on (Marshall, 1962). They have thus some means to visually recognize their surroundings, and some species, at least, have a territorial sense. Mid-water fishes, with no such habitual points of reference, seem capable only of producing incidental sounds, though these may still play some part in their lives, particularly if they are schooling kinds (Moulton, 1960). Most species of mesopelagic fishes have individual, and presumed recognition, patterns of light organs, which may be further elaborated in numerous species to facilitate intersexual activities. Yet, mesopelagic fishes have no evident sonic mechanisms, and the same is true of bathypelagic species. Most of these bathypelagic fishes have evolved macrosmatic males that are smaller than the microsmatic, scent-producing females. Besides using these olfactory signals, ceratioid males may recognize the light lures of their proper partners. But over the twilight depths of the continental slope, the trend is decidedly toward sonic sexual dimorphism in benthopelagic teleosts. Relatives of these sonic species from deeper reaches, where continuous vision is no longer possible, have no means of sound making. It looks as though sonic and visual stimuli reinforce one another in sexual behavior and even in territorial activities. Underwater craft and television should help us to test these ideas.

COVER AND CAMOUFLAGE

Conditions of life in the sea and a forest are curiously similar. Both these living spaces "have a vertical gradation in light, with associated gradations in other aspects of the environment, such as temperatures and air or water movement. This is due in one case to the arrangement of vegetation and in the other case to the properties of water itself. And both environments are relatively stable, in one case because of the insulating effect of the mass of vegetation and in the other because of the density and heat conservation of water. The physical properties of water underlie the insulation and stratification in both environments—though in one the water is free and in the other tied up in the protoplasm and sap of the forest trees" (Bates, 1960). We may also remember that some animals of both environments migrate diurnally and seasonally towards and away from the photosynthetic regions. And, corresponding to the many kinds of small animals that live in the leaf litter of the forest floor, there are quite diverse infaunas of deposit dwellers on the ocean floor.

Of all the biological contrasts between arboreal and marine environments, that concerning cover is one of the most outstanding. "Go for a walk in the woods with a naturalist: although you will assuredly see some animals, even conspicuous ones—let us say meadow brown butterflies, a green woodpecker, a swarm of flies dancing close over a stream, or bumble-bees visiting flowers—any idea that you are really seeing the animal population as a whole is quickly found to be an illusion. The caterpillar of the butterfly (which itself may be invisible on a cold, windy day) lives down among grass by day, climbing higher to feed at night; the woodpecker (which in any case is only visible now and then) subsists by picking out its food from dead wood or under bark, or inside mounds of the yellow ant which is entirely subterranean in habits; the larvae of the flies may live hidden in damp soil or under water, while the adults may be swarming for a few days in the year, and the bees are probably storing honey and pollen in an underground nest whither they retire at night. This huge world of animal populations is mostly hidden from ordinary observations by a *curtain of natural cover*: it begins to be fully revealed only when you look on the underside of leaves of a beech or a sycamore tree, strip the loose bark from a dead log or dig into its rotting sapwood, turn over a stone or a dead rabbit, or the litter of dying leaves—or go out at night with a mothing lamp. And there are many hundreds of animals that cannot easily be detected at all or properly counted without special methods of extracting them from the soil or other medium in which they live" (Elton, 1966, pp. 23–24). In the deep ocean there is, of course, the cover of twilight and darkness. The only tangible cover is mostly on the

Deep-Sea Fishes

bottom—in the shape of sediments, rocks, corals, sponges, ascidians, and so forth. In mid-water some shelter is provided by certain medusae and tunicates. Such benthic and mid-water cover is occupied by invertebrates, but not, at least, by adult fishes.

Before considering their cover, or rather covering, we should remember that deep-sea fishes develop little in the form of spiny or armored protection. Though spiny-finned fishes dominate shallow sea faunas, they are in a minority in the deep ocean. Even spiny-finned species, such as melamphaids and chiasmodontids, bear rather feeble spiny rays. Thus, lacking tangible cover and imposing armament, it is not surprising that deep-sea fishes have "had to turn" to concealment through their coloration and certain light-organ patterns. Here we shall consider the mid-water species.

A fish in aquatic space may merge with its radiant environment by being transparent to light, by reflecting light to match the background perceived by an observer, or in certain surroundings, by having a very low reflectance. The larvae of mesopelagic and bathypelagic fishes, which live in the photic zone, are largely transparent to light. And, most probably, the uveal tracts of their eyes, like those of other larval teleosts, develop an outer silvery layer (argenteum). This layer certainly conceals the underlying black pigment of the uvea, and it presumably reflects light so that the eyes tend to become invisible to a predator. During and after metamorphosis, when the mature coloration is developing, the young fishes are moving down to the dim or dark depths of their adult living space. The tail, which is usually the last part of the body to become pigmented, may still be transparent. Even in some adult fishes, such as certain paralepidids, the tail remains transparent or is at most translucent.

For many mid-water fishes their adult coloration is a uniform dark brown or black. Nearly every member of the bathypelagic fauna has this color, and so have about half the mesopelagic species, notably those belonging to the families Searsidae, Astronesthidae, Melanostomiatidae, Malacosteidae, Melamphaidae, and Chiasmodontidae. Some of the myctophids, bathylagids, and trichiuroids are also dark fishes. In dim or dark surroundings, as the black-faced members of night patrols are well aware, dark colors conceal very well. Moreover, dark brown or black skins reflect very little luminescent

61

light. Dark-skinned, deep-sea fishes thus enjoy good cover, and it is surely not by chance that most of them are predatory kinds.

Silvery fishes live above the threshold of light at mesopelagic levels. More precisely, such fishes have silvery flanks, usually a silvery iris, and a dark back. Of the gonostomatids, species of *Vinciguerria, Maurolicus, Ichthyococcus, Valenciennellus, Bonapartia, Polymetme,* and *Photichthys* are silvery-sided. Hatchet-fishes (Sternoptychidae) are classic instances of silvery fishes and so are the slender-tailed lantern-fishes (*Gonichthys, Tarletonbeania, Loweina,* and *Centrobranchus*). Most species of the lantern-fish genera *Myctophum, Benthosema,* and *Hygophum* and some kinds of *Electrona* and *Diaphus* also have mirrored flanks. Of the other iniomous suborder Alepisauroidea, species of *Alepisaurus, Anotopterus, Omosudis, Paralepis,* and *Notolepis* have silvery or brassy sides, usually with a marked iridescent sheen. Silvery members of the suborder Argentinoidea include species of *Opisthoproctus, Nansenia,* and *Leuroglossus.*

Lastly, but not completing a full survey,[15] *Gigantura, Lyconus, Stylephorus, Diretmus,* and *Lepidopus* are silvery fishes. Of some 750 species of mesopelagic fishes, more than a hundred have developed mirrored flanks.

There is a final group of pale to tan-colored mesopelagic fishes. *Cyclothone signata, C. braueri,* and *C. alba* fit here; so do many of the paralepidids other than *Paralepis* and *Notolepis.* In this group the darkest parts of the body are along the back and the abdominal flanks (where the black peritoneum may show through the thin lateral muscles). Other parts of the body tend to be translucent or even transparent.

The reflection of light from silvery-sided fishes has been studied by Denton and Nicol (1965, 1966). Under the surface waters—and whatever the altitude of the sun— the field of light is symmetrical about a vertical line and most intense in a downward direction. This field, though exponentially reduced, of course, is maintained with increasing depth, so that at mesopelagic levels, as early tests showed, the downward component is still the strongest. Evolving in such fields, fishes have "discovered" pigmentary means of vanishing from view.

15. Actually a full survey is not possible, for some mesopelagic fishes described as having grayish or dark flanks are probably silvery in life.

Deep-Sea Fishes

A fish "will be invisible from a given direction if the light reaching the eye of the observer is the same whether the fish is present or not" (Denton and Nicol, 1965). Ideally, the dark back should reflect enough of the strong down-going light to match the dim upward component of the ambient light, while the flanks should perfectly mirror the background from all relevant angles of view. Denton and Nicol have shown that fishes such as the herring and bleak go far to meet these ideal requirements, particularly in the elaboration of vertical mirrors. Silvery mesopelagic fishes doubtless have parallel powers of reflection. For instance, the extent of the dark dorsal pigment and the orientation of the reflecting platelets in the skin of a hatchet-fish and a *Stomias* are very like those of silvery fishes from shallow waters (Nicol, 1967; Figure 27).

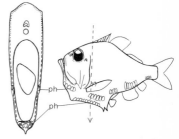

27 Silvery pigmentation of a hatchet-fish (*Argyropelecus*). Left: a transverse section (redrawn from Nicol, 1967) cut through the arrowed region, showing the extent of the dark and silvery pigments, the former shown as a broken line, the latter as dots. *ph*, photophore.

The evolution of dark and silver bodies by so many mesopelagic fishes gives some impression of selection pressures in this environment. Actually, most kinds of silvery fishes live at upper mesopelagic levels, where to the sensitive eyes of a predatory fish or squid, uncamouflaged prey will stand out against the background of light, except when viewed from above. We must remember, too, that most of these fishes have small mouths and that they feed on zooplankton organisms. Indeed, diverse kinds spend several hours at night hunting for their food in the upper waters, where frequent flashes of luminescence must sometimes expose their mirrored sides. Their black-skinned predators are not so exposed. Even so, plankton-eating fishes are more numerous than their enemies. Furthermore, by day at least, hatchet-fishes and lantern-fishes may live together in schools. A school of uncamouflaged fishes, as Denton and Nicol (1966) argue, is so much easier a target for a predator.

But silvery-sided fishes are very vulnerable from below. When a keen-eyed fish or squid looks upwards he will see their underparts, which are not covered by their mirrored flanks. The same, of course, is true of unmirrored fishes. Realizing this, Fraser (1962) has suggested, and Clarke (1963) has argued in more detail, that one function of the ventral photophores is to match the background of down-welling light. Since mesopelagic fishes tend to hug a particular isolume, a steady and apt glowing of their ventral lights will thus break up their silhouette. But, if such shadow elimination is to give a

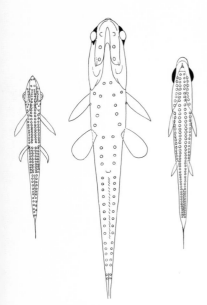

28 Light organs on the underside of three mesopelagic fishes. Left: *Vinciguerria*. Middle: a lantern-fish, *Myctophum nitidulum*. Right: *Ichthyococcus*.

fish fair protection from underlying predators, we might also expect the ventral photophores to shine continuously, at least during the hours of daylight. There is no evidence for this, though Nicol (1967) points out that luminescent light is weak and could be used to counteract ventral shadow over an intensity range of 3×10^{-10} to 5 to $10^{-4} \mu W/cm^2$. The first figure is Clarke and Denton's (1962) estimate of the threshold of vision for keen-sighted mesopelagic fishes.

This is an attractive hypothesis and it can also be applied to luminescent squids and crustaceans. With regard to fishes, there seem to be certain difficulties, which may be seen by comparing the underparts of a *Vinciguerria*, an *Ichthyococcus*, and a lantern-fish (Figure 28). An *Ichthyococcus* lives in dimmer waters than *Vinciguerria* spp., but its underparts are just as thickly studded with light organs. *Vinciguerria* species live at much the same levels as some lantern-fishes but have a better coverage of ventral lights. The tail of the lantern-fish though much narrower than its head and trunk, carries more potential light power per unit area of ventral silhouette. Even so, these objections depend on an all or nothing action of the ventral lights. Perhaps shadow elimination depends on the display of a selected constellation of lights. It will be interesting to see if this is true. If it is true, we should not expect silhouette removal to be the only function of ventral photophores. Black species of *Cyclothone*, which live below the threshold of light, retain their ventral photophores. And the singing midshipman, *Porichthys*, a bottom-dwelling fish, has a magnificent series of ventral lights.[16]

VERTICAL MIGRATIONS

Under the cover of darkness, many mesopelagic fishes make daily visits to the upper waters of the ocean. Before sunrise they move down to their daytime living spaces. These diurnal migrations run parallel to those made by many members of the zooplankton, micronekton, and nekton. Indeed, such is this nightly, near-surface con-

16. Observations from deep-sea submersibles suggest more and more that some luminous mesopelagic fishes (and the nonluminous ones) are quiescent by day, when they are often orientated away from the horizontal plane.

centration of diverse animals that both plankton-feeding and predatory fishes move up to take advantage of this daily enrichment of their levels in the food pyramid. Whatever may be the full biological significance of these movements, most of the mesopelagic fishes habitually seek their food near the primary source of life in the ocean. There is no evidence that they are making spawning migrations.

The vertical movements of fishes have been seen by observers in bathyscaphes and followed by echo sounders. But until echo traces can be correlated with the species that cause them, the only way of studying migrations is by using a well-planned series of closing nets. For a detailed survey we must await the results of recent "Discovery" cruises, though much has been done by Pearcy and Laurs (1966) off Oregon. Towing between depths of 0–150, 150–500, and 500–1,000 meters, they found evidence for migrations (and net avoidance) above a level of 500 meters, mainly for three myctophids and a melanostomiatid. There were no evident upward migrations from levels between 500 and 1,000 meters, a conclusion supported by echograms taken during the investigations. Indeed, deep scattering layers are rarely recorded below 700 meters. The vertical migrations of deep-sea fishes, so far as we know, are confined to mesopelagic species.

Most diurnal migrators belong to the families Myctophidae, Gonostomatidae (mainly the smaller species), Sternoptychidae, Astronesthidae, Melanostomiatidae, Stomiatidae, and Chauliodontidae. Certain of the trichiuroids (for example, *Gempylus* and *Nealotus*) also appear in the surface waters at night. Members of the first three families are predominantly consumers of zooplankton organisms: those belonging to the others are predacious fishes, depending mostly on animals of the micronekton and nekton. Some species, notably hatchet-fishes, rarely if ever migrate upwards beyond the "seasonal" thermocline, where some accumulation of animals may be expected. If the pale species of *Cyclothone* and the melamphaids carry out vertical migrations, they have yet to be detected.

Most kinds of deep-sea fishes live between the north and south subtropical convergences, below the warmer waters of the ocean. Thus, during their upward migrations most mesopelagic fishes will experience some rise in temperature. If 400 meters is taken as a mean starting level, migrants between latitudes 40°N and 40°S will be warmer by 5 to 10°C or even more, by the time they reach the surface waters. Such changes would harm many shallow water species that live at temperatures equal to those met at the daytime levels of migrating mesopelagic fishes. In a study of migrating zooplankton organisms, McLaren (1963) has followed the implications of a mathematical model based on von Bertalanffy's growth equations. The model predicts that a migration from cool to warmer waters would be advantageous. An animal that is more active and gets all its food in the surface waters, then retires to rest by day in cooler underlying waters, will gain an energy bonus that can be put into fecundity. Moreover, fecundity will be greater, for during its daytime resting periods an animal will grow larger (though more slowly) than it would in warmer surroundings. There is some evidence bearing on this hypothesis with regard to mesopelagic fishes, for they have actually been seen at rest by day. Furthermore, though species of *Vinciguerria* and *Maurolicus* do make vertical migrations in the isothermal Red Sea, they are smaller, and presumably less fecund than closely related forms that live at lower temperatures in the Indian Ocean. There might also be another advantage in "resting" by day. The more a fish moves, the more readily it is sensed by its enemies.

ORGANIZATION AND ENVIRONMENT

We have now seen some of the basic differences in the organization of mesopelagic, bathypelagic, benthopelagic, and benthic fishes of the deep ocean. The present endeavor will be to integrate such contrasts by considering their correlations with the environment, particularly with the living components. I have now found more to add to an earlier survey (Marshall, 1960).

A paradigm of the contrasting organization of mesopelagic and bathypelagic fishes is provided by two species of the genus *Gonostoma*. The first, *G. denudatum*, lives at mesopelagic levels in the Mediterranean and Eastern Atlantic. Its relative *G. bathyphilum*, swims in bathypelagic reaches below 1,000 meters in the Atlantic Ocean (Figure 29).

G. denudatum, which grows to a standard length of at

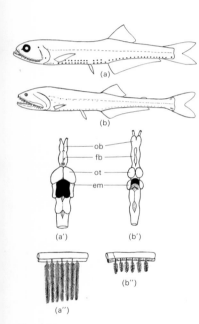

29   A comparison of two related fishes. (a) *Gonostoma denudatum*, from the mesopelagic zone, and (b) *Gonostoma bathyphilum*, from the bathypelagic zone. Note the much smaller eyes and light organs of the latter. Below are shown (drawn to the same scale) their brains and a few of the longest gill filaments on the first gill arch; *denudatum* (*a'* and *a''*); *bathyphilum* (*b'* and *b''*). In the brain the corpus cerebellum is shown black. *ob*, olfactory bulb; *fb*, forebrain; *ot*, optic tectum; *em*, eminentia granularis.

least 300 mm, is a dark-backed fish with silvery, compressed sides and a full set of prominent light organs. The
suspensorium, set back from the vertical axis at an angle
of about 30°, bears strong and relatively long lower jaws.
Both jaws hold numerous pointed teeth, some longer
than the others. The eyes and olfactory organs are
moderately large, the former with well-proportioned
dioptric parts, the latter of comparable size and structure in both sexes. Matching the development of these
sense organs, the optic tectum and forebrain are comparatively large; so is the corpus cerebellum (Figure 29).
We are thus not surprised to discover a well-knit series
of lateral muscles, for in fishes there is a direct correlation between the development of myotomes and corpus
cerebellum. In turn, muscles move a light but well-
ossified skeleton, including scales that cover the trunk
and tail. Muscles and skeleton, the two heaviest major
parts of the fish body—and the other organs—are
carried at neutral buoyancy by a capacious swimbladder
with a large gas gland and rete mirabile (Marshall, 1960).

Gonostoma bathyphilum is entirely black and its photophores are regressed. It has longer jaws than G. denudatum, carried on a suspensorium swung at an angle of 50°
to the vertical axis. The eyes, which are no more than half
the diameter of denudatums, are ill-proportioned dioptrically and probably regressed. In female fishes, and they
attain a length of at least 200 mm, the olfactory organs
are reduced. The smaller males, growing to some 150 mm
have large olfactory organs (Marshall, 1967). Except for
the well-developed forebrain of the males, the parts of the
brain in bathyphilum are much smaller than those of
denudatum (Figure 29). The poorly developed corpus
cerebellum is reflected in the rather weak, loosely knit
lateral muscles, which are particularly reduced in the tail,
compared to those of G. denudatum. Moreover, the
muscles move a poorly ossified skeleton. There is no
swimbladder, but G. bathyphilum, largely because of its
reduced muscles and light skeleton, is probably close to
being neutrally buoyant.

It will now be clear that Gonostoma denudatum has the
more complex organization. Consider the contrasts
between the two species in a little more detail. First, the
only structures that are larger in bathyphilum are the jaws
and suspensoria, which probably means that this species
is able to take rather large prey. The complex retinae of

denudatum not only require complex centers in the brain but also elaborate choroidal "glands" to supply oxygen to the visual and ganglion cells. *G. bathyphilum* "saves" these ocular expenses. The closely knit muscles of *denudatum* need more nervous and vascular elaboration—as well as a firmer skeleton and more cerebellar coordination—than those of its relative. *G. bathyphilum* also "saves" the expense of developing and maintaining the vascular, glandular, and muscular parts of a swimbladder.

Indeed, so relaxed are the above systems of *G. bathyphilum* that we would expect it to have smaller and much less elaborate gills than its relative (Figure 29). The gill surface per unit of weight in *denudatum* is probably more than five times the equivalent figure for *bathyphilum*. Again, the heart of the well-found species is at least three times the volume of the other one. Lastly, the kidneys of *denudatum* are larger and contain many more tubules than those of *bathyphilum*.

It is thus fair to infer that *denudatum* is the more active species and most likely it undertakes diurnal migrations. There is no evidence that *bathyphilum* is a migrator. In all probability, it is a hovering and darting kind of predator. This is suggested, at least by its extended dorsal fin, which partly opposes the anal fin to form something like the fin feathering found on pike-like predators. Moreover, like the pike, female *G. bathyphilum* have few red muscle fibers in their myotomes. Both sexes of its more active relative have a well-formed outer layer of red fibres: they are built like a "stayer."

These outlines of the organization and life of two related deep-sea fishes will serve as a guide to further comparison of mesopelagic and bathypelagic species. Virtually all the above contrasts will be found in small mesopelagic gonostomatids, such as *Vinciguerria*, *Maurolicus*, and *Bonapartia* and in their relatives, the black bathypelagic species of *Cyclothone*. Some of the contrasts, such as those between muscle, brain, and gill systems, are even more striking (Marshall, 1960).

We may now widen still further the comparison between mesopelagic and bathypelagic fishes. Considered broadly, there are three groups of mesopelagic fishes: small-jawed plankton eaters, most of which have a swimbladder; predators with a swimbladder; and predators without a swimbladder.

Deep-Sea Fishes

The most diverse members of the first group are small gonostomatids, hatchet-fishes, argentinoids, lantern-fishes, and melamphaids. Such species are compared in the table below with bathypelagic fishes: *Cyclothone* spp., ceratioid anglerfishes, gulper eels, and so on.

Organization of mesopelagic and bathypelagic fishes

Features	Mesopelagic, plankton-consuming species	Bathypelagic species
Color	Many with silvery sides	Black
Photophores	Numerous and well developed in most species	Small or regressed in gonostomatids; a single luminous lure on the females of most ceratioids
Jaws	Relatively short	Relatively long
Eyes	Fairly large to very large, with relatively large dioptric parts and sensitive pure-rod retinae	Small or regressed, except in the males of some anglerfishes
Olfactory organs	Moderately developed in both sexes of most species	Regressed in females but large in males of *Cyclothone* spp. and ceratioids (most species)
Central nervous system	Well developed in all parts	Weakly developed, except for the acoustico-lateralis centers and the forebrain of macrosmatic males
Myotomes	Well developed	Weakly developed
Skeleton	Well ossified, including scales	Weakly ossified; scales usually absent
Swimbladder	Usually present, highly developed	Absent or regressed
Gill system	Gill filaments numerous, bearing very many lamellae	Gill filaments relatively few with a reduced lamellar surface
Kidneys	Relatively large with numerous tubules	Relatively small, with few tubules
Heart	Large	Small

Mesopelagic predators with a swimbladder include species of *Astronesthes*, trichiuroids, and the chiasmo-dontids. The contrasts between the organization of these fishes and that of bathypelagic species are also very marked. For a paradigm, consider the systems of *Astronesthes niger* and a female anglerfish of the species *Neoceratias spinifer*. Every tissue system of the former is elaborated to a far greater extent, much as we saw in *Gonostoma denudatum* by comparison with *G. bathyphilum*. For instance, the lateral muscles of *Astronesthes*

69

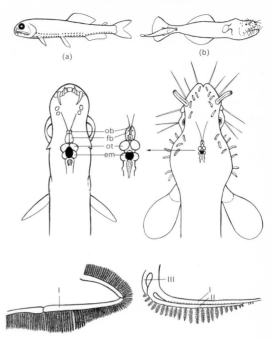

30 Comparison of two predatory deep-sea fishes drawn to the same scale. (a) *Astronesthes niger,* from the mesopelagic zone, and (b) a female anglerfish, *Neoceratias spinifer* (with attached male) from the bathypelagic zone. Below each fish is a drawing of the head and brain (× 2) and part of the gill system (× 3). The corpus cerebellum of the brain is shown black. Between the two heads the tiny brain of *Neoceratias* is shown enlarged (× 7). *ob,* olfactory bulb; *fb,* forebrain; *ot,* optic tectum; *em,* eminentia granularis. Note that the gill system of *Neoceratias* is very reduced. As in other ceratioids there are no filaments on the first gill arch.

are larger, more closely knit, and contain many more outer red fibers than do those of a *Neoceratias* of the same size.[17] This difference is reflected in the great contrast between the size of the corpus cerebellum in these two species (Figure 30). In fact, the corpus cerebellum of the *Astronesthes* is nearly equal in volume to the entire brain of the *Neoceratias*, every part of which, except for the acoustico-lateralis centers, mirrors the reduction of associated organs.

More generally, the main contrasts between the or-

17. Waterman (1948) wrote as follows about another female ceratioid: "Like the skeletal elements, many of the muscles in *Gigantactis* are reduced in size, and some commonly present in the less highly specialized actinopterygians are lacking altogether."

Deep-Sea Fishes

ganization of mesopelagic predators with a swimbladder and that of bathypelagic species are just as trenchant as those given in the above table for the latter fauna and mesopelagic plankton-eaters. If the mesopelagic predators were put in this table in place of the plankton eaters the main descriptive changes needed would be those correlated with their predatory life: black skin, large jaws and teeth, and so forth.

Mesopelagic predators without a swimbladder belong mainly to the stomiatoids (Melanostomiatidae, Stomiatidae, Chauliodontidae, Malacosteidae) and alepisauroids (Scopelarchidae, Evermannellidae, Omosudidae, Alepisauridae, Anotopteridae, and Paralepididae). Here a good guide to the contrasting organization of these and bathypelagic fishes is given by comparing *Gonostoma elongatum*, a representative of the first group, and *G. bathyphilum*. The first species, which grows to a standard length of 250 mm or more, feeds on fishes (for example, *Cyclothone*) prawns, euphausiids, and so on. If the reader will refer to the earlier comparison of *G. denudatum* and *G. bathyphilum*, the main descriptive changes needed to put *elongatum* in place of *denudatum* would be these: (1) *elongatum* is centered at lower, rather than upper, mesopelagic levels; (2) the coloration is black and brassy; (3) the eyes and optic centers of the brain are moderately developed; (4) the lateral muscles and skeleton are not so well formed as those of *denudatum*; (5) the swimbladder is regressed and invested by adipose tissue; (6) the gills are not so well developed.[18]

The main organ systems of *Gonostoma elongatum* have also been considered in relation to its buoyancy balance sheet (Denton and Marshall, 1958). Though this species is without a gas-filled swimbladder it is close to being neutrally buoyant, largely because of its lightly ossified skeleton and rather weakly developed myotomes (Figure 31). Even so, *elongatum* has better muscles than its relative *bathyphilum*. Not only are the lateral muscles larger in *elongatum*, but they also carry a substantial outer layer of red fibres, which are much sparser in females of *bathyphilum*.

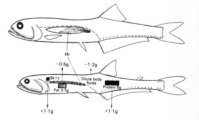

31 Comparison of the buoyancy relations of *Gonostoma atlanticum* (above) and *Gonostoma elongatum* (below). (After Denton and Marshall, 1958.) *G. atlanticum* has a gas-filled swimbladder (*sb*), but this is regressed in *G. elongatum*, which depends largely on its reduced skeletal and muscle systems for attaining a near buoyant condition.

18. On the first gill arch of a 120 mm *G. elongatum* I counted about 200 gill filaments, the longest measuring 2 mm, and there are about 30–35 gill lamellae per mm of length. Comparable figures for a 130 mm *G. denudatum* are 210 gill filaments, the longest 2.5 mm with about 50 lamellae per mm.

Again, if we make an overall comparison of such meso-pelagic predators and bathypelagic fishes, the same contrasts in organization already seen will appear once more. But certain of these contrasts are not so sharp as those found in a comparison of both groups of meso-pelagic fishes with a swimbladder and the bathypelagic species. The main lessening of contrast is in the weaker skeletal and muscular systems of the mesopelagic predators without a gas-filled swimbladder.

Bathypelagic fishes thus have a much simpler and more relaxed organization than the mesopelagic species. More-over, the regression or loss of the swimbladder is not related to high ambient hydrostatic pressures. In bentho-pelagic fishes living at depths well below 2,000 meters, this organ is highly developed (Marshall, 1960, 1964). But if mesopelagic fishes, such as *Gonostoma elongatum* and *Xenodermichthys copei*, come close to being neutrally buoyant, bathypelagic fishes—with even more reduced skeletal and muscular systems—may well be able to hover with virtually no expenditure of energy.

I have argued elsewhere (Marshall, 1960), that bathy-pelagic fishes are sparely organized to exist in a food-poor environment. Between levels of 1,000 and 3,000 meters the biomas of zooplankton is about one tenth to one hundredth of that in the surface layer (Figure 32). This is not surprising when we remember (1) that the biomass of deep-sea animals is related to productivity in the euphotic zone, which between the subropical conver-gences is not subject to much seasonal variation but is relatively low; and (2) that organic particles derived from this zone sink so slowly that they have lost most of their nutritional value to bacteria by the time they reach bathy-pelagic levels. We can understand that: "The standing stock of pelagic suspension feeders decreases rapidly with increasing depths, perhaps to disappear completely at about 3,000 m; or even less, since records of maximum depth of pelagic animals that also live at lower depths may be misleading" (Jørgensen, 1966, p. 293). Jørgensen believes that the biomass of deep, mid-water suspension feeders is more closely related to the sparse, ambient crops of heterotropic micro-organisms than to the (largely unusable) concentrations of organic particles. At all events, in descending to bathypelagic levels there is a marked increase in the numbers of predatory forms in the zooplankton. Many bathypelagic copepods, for

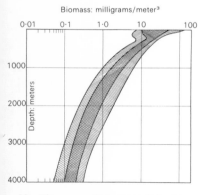

Biomass: milligrams/meter³

32 The range of biomass of plankton between 0 and 4,000 meters in the Pacific and Indian oceans. The range curve for the Pacific Ocean is cross-hatched upwards from left to right that for the Indian Ocean cross-hatched downwards from left to right. (Redrawn from Vinogradov, 1968.)

instance, are first-order carnivores. Thus, most of the food organisms of a *Cyclothone*, such as young fishes, euphausiids, arrow-worms, and copepods, are dependent themselves on the third or fourth level of the food pyramid. Bathypelagic fishes with larger mouths and stomachs, female ceratioids and gulper eels, will depend even more on the fourth level of the pyramid.

Besides evolving sparely organized, low-cost tissue systems, which also save energy through their overall neutral buoyancy, bathypelagic fishes have acquired energy-saving habits. Female angler-fishes (and perhaps gulper eels) lure their prey rather than chase it. *Cyclothone* species probably spend much of their time hanging motionless in the water, there to dart on any prey they may perceive. Yet for part of their life, male fish must have the stamina needed to find their mates. Indeed, how can the macrosmatic males of ceratioids, *Cyclothone* spp. and *Gonostoma bathyphilum* find their scent-producing partners if they do not swim persistently into currents on the receipt of the right olfactory clues? Even though dwarf male ceratioids easily outnumber the ripe females, it still seems remarkable that any of these small fishes (mostly from 20 to 40 mm long) are able to find a mate. But male ceratioids are much better shaped for effective swimming than their grotesque partners. Moreover, from head to tail fin there is a well-formed layer of red fibres over their myotomes. In males of *Linophryne* spp., I found the red muscle was 3 to 5 fibrers in thickness, and it extends over the chevron-shaped parts of the myotomes (Figure 33). Female ceratioids have at most a thin restricted layer of red fibres. Using the fat stored in the red muscle, and drawing on reserves in the large liver, we may suppose that the males have the fuel to cruise persistently and rheotactically in search of the source of a scent trail.

Male *Cyclothone* and *Gonostoma bathyphilum* also have more red muscle than the females (Figure 33). Again, the red fibres extend over all of the myotomes and cover the chevron-shaped sections. The convergences with the ceratioids need no more elaboration.

In ceratioid angler-fishes and black *Cyclothone* spp., the two dominant groups of bathypelagic fishes, the males are smaller than the females. Through the evolution of dwarfed males, not only is the energy budget of each population of a species reduced, but there is less competition for food between the sexes. Such competition

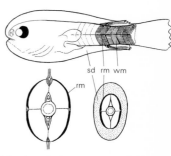

33 A male deep-sea anglerfish (*Linophryne sp.*) showing red muscle fibers (*rm*) overlying the white fibers (*wm*) of the muscle segments. Below, right: a transverse section through the tail showing (in black) the red muscle layer. Between the skin and the muscles is a large subdermal space (*sd*). Below, left: a transverse section through the tail of a male *Cyclothone livida*.

must be virtually absent in ceratioid angler-fishes. Indeed, Bertelsen (1951), who found that the liver of the males reaches its maximum size at metamorphosis, concludes that they do not feed after this stage of their life. Before metamorphosis, when both sexes have the same size and diet, they live in surface and mesopelagic waters, where there is more food.

Over oceanic regions the biomass of plankton at bathypelagic levels faintly reflects the overlying productivity of the euphotic zone. Naturally, this productivity is not everywhere the same. Thus, if the low-cost organ systems of bathypelagic fishes adapt them to food-poor environments, we should expect these systems to be most economical in regions of lowest productivity. The Mediterranean is certainly one such region, particularly over its eastern reaches. Ceratioid angler-fishes have never been taken in this sea: its bathypelagic fish fauna is evidently dominated by small black *Cyclothone*, which were studied by Jespersen and Tåning (1926). They considered these fishes to be a subspecies, *pygmaea*, of *C. microdon*. In the Atlantic, *microdon* grows to a standard length of some 70 mm, but its Mediterranean relative is a dwarf, attaining no more than 27 mm. Presumably, this is the maximum length of female *pygmaea*: males may well be somewhat less than 20 mm when fully grown. At all events, Jespersen and Tåning found about 1,000 eggs in the ovaries, whereas North Atlantic females of *microdon* carry some 10,000 eggs. This marked dwarfing and lowered fecundity in the Mediterranean is surely related to the low productivity of this sea. Certain other Mediterranean fishes, though to a lesser extent, are also smaller than those of the same species from the Atlantic. Indeed, examples from the mesopelagic fauna will be considered at a later stage. Meanwhile, further observations of a bathypelagic *Cyclothone* are relevant here. This species, *acclinidens*, is distributed in part across the equatorial belt of the Pacific Ocean, where productivity is greater over the eastern half. The standing stocks of *Cyclothone* food should thus be greater in the east, and it is here that Mukhacheva (1966) found *C. acclinidens* to have relatively long and free gill filaments. In western individuals, on the other hand, the filaments are half as long and they fuse into a narrow band along the cerato and hyobranchial elements of the gill arches. In view of our earlier observations of the close connection between the overall level of organization of deep-sea fishes and the degree of development of their gills, Mukhacheva's findings acquire an added significance.

The kinds of bathypelagic fishes that are derived from benthopelagic ancestors are represented by the macrourids *Squalogadus*, *Macrourides*, *Echinomacrurus*, *Cynomacrurus*, and *Odontomacrurus*[19] and in all probability by the aphyonine group of brotulids (Marshall, 1960, 1965). The benthopelagic relatives of these fishes are highly organized indeed when compared to any kind of primary bathypelagic fish. Benthopelagic macrourids and brotulids contain a capacious swimbladder, in which the length of the retia mirabilia is directly related to the depths of their living spaces. Having this float, they are able, as photographs and films show very well, to hover easily over the deep-sea floor. As they do so, their swimbladder is supporting a firm skeleton and a long series of well-knit myotomes, to consider only the heaviest tissues, with little metabolic cost. But the bathypelagic macrourids and brotulids have had to imitate the organization of the primary dwellers of this living space. To gain an impression of the changes needed to turn a benthopelagic macrourid into a bathypelagic one, compare *Trachyrhynchus* with *Squalogadus* (using individuals with the same length of head). *Squalogadus* resembles a tadpole. The inflation of the head, which develops also in *Echinomacrurus* and *Macrouroides*, is largely due to the extremely capacious, mucous-charged canals of the lateralis system. The trunk and tail are relatively short, so limiting the mass of muscle. Moreover, the skeletal parts, particularly of the head, are lightly built: so are the scales. A *Trachyrhynchus* has a greater mass of lateral muscles, a heavier skeleton, and formidable scales. But

19. *Odontomacrurus* and *Cynomacrurus* are known only from mid-water nets, but representatives of the first three genera have also been taken in bottom trawls. Concerning *Squalogadus* and *Macrouroides*, which are very closely related, the first is known mainly from bottom trawls, but the second, though represented by fewer specimens, has mostly been taken in mid-water nets (Marshall and Tåning, 1966). Bottom trawls may, of course, catch fishes as they are hauled to the surface. On the other hand, mesopelagic fishes, such as myctophids, have been seen near the bottom (Marshall, 1960), and at least one bathypelagic fish *Eurypharynx pelecanoides* was found to contain the remains of a sea urchin. But the convergences of these macrouroids with the primary inhabitants of bathypelagic regions, combined with the evidence from nets, indicates their bathypelagic bias.

it has a well-formed swimbladder, which is regressed in *Squalogadus*. In all probability the reduced muscles and skeleton bring *Squalogadus* close to neutral buoyancy. And, as we saw earlier, the developments of lateral muscles and the corpus cerebellum are directly correlated. The corpus of a *Squalogadus* is no more than one fifth the volume of its homologue in *Trachyrhynchus*. Lastly, the relatively small area of gill surface in *Squalogadus* reflects its low-cost tissue systems. Much the same contrasts would emerge in a comparison of other bathypelagic macrourids to their benthopelagic relatives. Again, the same would be true of a comparative study of aphyonine brotulids and the benthopelagic forms (Marshall, 1960). To imitate primary bathypelagic fishes, these secondary inhabitants have thus had to change drastically the organization they first inherited from their benthopelagic ancestors.

How are mesopelagic fishes able to maintain much higher levels of organization? The biomass of suitable food organisms at, say 500 meters, is likely to be at least 5 times the amount one would get at any bathypelagic level. More significantly, mesopelagic fishes live relatively close beneath the productive euphotic zone. In this zone, along with the resident populations of animals, consisting largely of herbivores, there is a nightly expansion of the potential food pyramids of mesopelagic fishes through the vertical migrations of copepods, euphausiids, prawns, arrow-worms, squid, fishes, and so forth. Echograms of deep scattering layers indicate that mesopelagic fishes reach these surface feeding grounds after swimming upwards for one hour or more.

An ascent of several hundred meters seems a rather strenuous undertaking. But Hutchinson (1966), studying freshwater zooplankton, shows that the metabolic cost is quite low. If an organism has an excess density of 0.015 a force of 14.7×10^{-8} dyne must be overcome per milligram of biomass during its ascent. "For a migration path of 50 m or 5,000 cm the energy needed will be 73.5 ergs or 1.75×10^{-6} g-cal. Since 1 mg of glucose in solution yields on oxidation 699 g-cal the minimum amount of glucose that would have to be oxidized to produce this energy would be 4.5×10^{-7} mg. Since we may reasonably assume that an organism of mass 1 mg and of excess density 0.015 would contain at least 1.2 per cent or 12×10^{-3} mg of organic matter, it is evident, even assuming an efficiency as low as 1%, less than one two-hundredths of the organic matter of the body would be oxidized in performing the ascent".

The metabolic advantages of vertical migration for micronektonic forms have been studied recently by Teal and Carey (1967). Bearing MacLaren's hypothesis in mind, they calculated the metabolic gain for epipelagic euphausiids that migrate to the surface waters at night. For instance, if *Euphausia hemigibba* "spend 12 hr. at 500 m and the other 12 hr. migrating and feeding at the surface, they would consume 15.1 ml O_2/g/day or save an amount equivalent to the consumption of 7.7 ml O_2/g/day." For *Thysanopoda monacantha*, a mesopelagic euphausiid that may not migrate up as far as the euphotic zone, they predict a metabolic loss. The unknown factor is the amount of food consumed by this euphausiid during and after its migration. Below the seasonal thermocline there may often be rich or concentrations of suitable food.

For migrating mesopelagic fishes weighing a few grams, which are nearly or just, neutrally buoyant, the metabolic cost is related mostly to the requirements of red muscle in moving the fish upward. Moreover, little energy would be needed during the descent when all mesopelagic fishes, whether they have a swimbladder or not, are likely to be negatively buoyant (Marshall, 1960). The entire energy put into vertical migrations and the capture of food would evidently not yield so much potential energy over the same time had the fish stayed at mesopelagic levels, where the biomass of zooplankton is smaller. Even "a relatively small advantage conferred by vertical migration would in fact make that process an economically significant one, favoured by natural selection" (Hutchinson, 1966).

Moreover, as I implied in the preceding paragraph, mesopelagic fishes have the right kind of muscular organization for sustained efforts. Their myotomes contain many red muscle fibres as well as white. Red fibres, which are usually formed on the outside of the myotomes, are finer than the white kind. They contain fat, much glycogen, myoglobin, many mitochondria and they are richly supplied with blood. White fibres hold little or no fat, relatively little glycogen, no myoglobin, few mitochondria and are more sparely fed with blood. These distinctions between the two sets of fibres have been known for

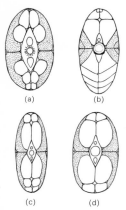

34 Transverse sections through the tail of four mesopelagic fishes showing the extent of the red muscles (stippled). (a) *Notolepis coatsi*, an Antarctic paralepidid; (b) *Electrona antarctica*, a lantern-fish; (c) *Maurolicus*, a gonostomatid; (d) *Astronesthes lucifer*.

some time, but for their functional contrasts we owe much to the recent work of Barets (1961) and Bone (1966). Bone, who studied the dogfish, *Scyliorhinus caniculus*, discovered from electrical recordings that the red fibres alone are active when the fish swim slowly, whereas during vigorous movements the white fibres are brought into play. Bone concluded that the slow, red parts of the myotomes, which apparently work through the oxidation of fats, are used when a fish cruises. The fast, white fibres serve "as an emergency store for vigorous movements." The red muscle will go on working for hours, but the white muscle soon becomes fatigued.

Before these investigations, Boddeke, Slijper, and van der Stelt (1959) had correlated the muscle patterns of European freshwater fishes with their mode of life. Pike, perch, sticklebacks, river bullhead, and stone loach, which dart on their prey, have very little red muscle in their myotomes. Such muscle is better developed in species such as dace, ide, roach, rudd, bream, gudgeon, and tench, which in feeding "must develop a small amount of energy during a comparatively long time." Active species with much stamina, such as salmon and trout, not only have a substantial layer of red muscle fibers around their myotomes, but some red fibres are mixed with the inner, white muscle.

In a series of mesopelagic fishes, I found that the relative development of red muscle equals or even exceeds that found in the second group of freshwater fishes. The species dissected and examined microscopically are either largely dependent on zooplankton (*Vinciguerria, Maurolicus, Sternoptyx, Argyropelecus, Bathylagus, Searsia, Electrona, Myctophum, Lampanyctus,* and *Melamphaes*) or on micronekton and nekton (*Astronesthes, Chauliodus, Stomias, Eustomias, Odontostomias, Heterophotus, Bathyphilus, Malacosteus, Notolepis, Evermannella, Nealotus,* and *Promethichthys*). The comparative development of red muscle in the tail of selected species may be seen in Figure 34.

Our earlier review of the variable elaboration and functions of red muscle suggests that mesopelagic fishes turn to this muscle for sustained effort, as when they are migrating or searching for food. Their red muscle enables them to take advantage of the near-surface feeding grounds. Thus, they have the means to support far higher levels of organization than the bathypelagic fishes. After all, species with a swimbladder automatically carry a firm skeleton and well-formed myotomes at neutral buoyancy. Predatory forms without a swimbladder (such as *Chauliodus, Stomias, Eustomias, Odontostomias, Heterophotus,* and *Bathyphilus*), though coming close to neutral buoyancy by some reduction in their myotomes and skeleton, still develop a relatively large red muscle system. An outstanding instance is *Notolepis coatsi,* an antarctic paralepidid, which appears in the surface waters at night and feeds on krill (*Euphausia superba*), small fishes, and so on. Figure 34 shows that about half the cross section of tail muscle consists of red fibres, which are presumably active when *Notolepis* is ranging upward in search of food. In a predatory trichiuroid, *Nealotus tripes,* the corresponding fraction is about one sixth. Thus though *Notolepis, Nealotus,* and other voracious mesopelagic fishes, such as melanostomiatids and malacosteids, are reminiscent of pike, the convergence does not extend to their muscular organization.

To recapitulate, primary productivity, which is reflected in the standing crops of zooplankton at mid-water levels, varies from one oceanic region to another. Just as at bathypelagic levels, then, the biomass of zooplankton in mesopelagic waters of the Mediterranean is considerably lower than the amounts at corresponding depths in the outer Atlantic. Now, *Cyclothone braueri,* a pale-colored mesopelagic species, grows to 30 mm in the Mediterranean, but reaches some 45 mm in the North Atlantic. This dwarfing of fishes that live below waters of low primary productivity is found also in those that inhabit the Central Water Masses. These water masses are formed by the sinking of water at the subtropical convergences and they overlie the axis of the permanent thermocline. Turning slowly above the Central Water Masses, and tracing their expanse, are the great gyres of the oceans, where the water is a deep blue. In the gyres the near-surface thermocline is so stable that it markedly limits the refreshing of the euphotic zone with nutrient salts from underlying waters. Plant productivity is thus low in comparison, for instance, to that maintained by upwelling along the belts of divergence in the equatorial current systems. Few birds and school of fishes find a living in these poor subtropical waters. Standing crops of zooplankton are consequently low in the underlying

Central Water Masses. During the daytime, deep scattering layers, of which mesopelagic fishes with swimbladders are prominent components, are sometimes absent from the records of echo-sounders (Tchindonova, 1966).

Is it surprising that mesopelagic fishes of the Central Water Masses tend to be small? For instance, Ebeling (1962) found that four dwarf species of *Melamphaes*, which grow to an inch or less in length, "occur exclusively in the relatively sterile western equatorial and central water masses." Larger (4 to 6 inch) species of *Melamphaes* live in waters of higher productivity, such as those on the periphery of the Central Water Masses. In the North Atlantic, *Chauliodus danae*, which inhabits the Central Water Mass and underlying waters, grows to about half the size of *C. sloani*, (300 mm), found in richer peripheral waters. But these and other mesopelagic fishes of the Central Water Masses still maintain higher levels of organization than the bathypelagic species that swim below them.

If a benthopelagic fish is to invade bathypelagic regions, its organization must evolve towards a markedly lower level. It seems most likely that the bathypelagic macrourids and brotulids were edged out of their former benthopelagic existence, for there is more food near and on the deep-sea floor than at bathypelagic depths. The mean biomass of benthic organisms from 200 to 3,000 meters in the world ocean is about 20 grams per square meter (Vinogradova, 1962). Between the same levels in oceanic mid-waters, the mean biomass of zooplankton per cubic meter is certainly not more than one hundredth of the benthic figure. Moreover, present estimates of benthic biomass do not fully include the diverse fauna of small organisms, particularly polychaets, that live in the oozes (and are only retained by fine sieves). Benthopelagic fishes can also feed on the many organisms, notably crustaceans, that swim near the deep-sea floor. Recent exploration with an epibenthic sledge net has shown how diverse and numerous these organisms can be, even at depths beyond 4,000 meters (Hessler and Sanders, 1967).

The standing stocks of deep-sea benthos are greatest over the continental slopes, which receive much organic matter derived from the land and coastal waters. Even at depths between 1,000 and 3,000 meters which are also no

more than 100 miles from the nearest coastline, the biomass of benthos is generally above 1.5 g/m² (Vinogradova 1962). In general, quantities of benthic organisms fall as distance from the land increases and are
lowest under the gyres of the subtropical ocean, where we
saw that primary productivity is low. In the Pacific
Ocean, for instance, this impoverished biomass varies
between 0.1 and 0.05 g/m² (Filatova and Zenkevitch,
1966).

But there is no sharp contrast in organization, such as
was found for mesopelagic and bathypelagic fishes, between shallower and deeper dwelling benthopelagic
fishes. Except for their small or regressed eyes and optic
centers, macrourids and brotulids that live at depths
beyond 1,000 meters develop essentially the same levels
of organization as their relatives from lesser depths. As
the depth increases, the main problem with retaining a
swimbladder, which most benthopelagic fishes do, is not
so much the steepening pressure gradient opposing the
secretion of gases (see Kuhn, *et al.*, 1963) but rather the
decreasing buoyancy of the swimbladder. At 5,000
meters, the specific gravity of swimbladder gases, consisting mainly of oxygen, is about half the value at 500
meters. Perhaps this is why there seems to be some
reduction in the ossification and myotomes of such
deep-living macrourids as *Chalinura* and *Lionurus*
(Marshall, 1960). The benthopelagic (and benthic) fishes
that live below the impoverished subtropical waters,
must have the hardest struggle to make a living. Evidently
there is a very sparse fauna of fishes in these regions,
though further investigations are needed.

Photographs of benthopelagic fishes often show them
heading into currents, the direction of which is revealed
by the bending of such sessile invertebrates as sea lilies
and gorgonians. Even at depths beyond 3,000 meters,
current velocities of 6 to 15 cm/sec. have been recorded
(Hollister and Heezen, 1966). In facing currents and
ranging over the deep-sea floor in search of food, benthopelagic fishes would seem to need muscles capable of sustained effort. There is, in fact, a median strip of red
muscle along the tail myotomes of macrourids, brotulids, halosaurs, and so on. Moreover, the muscle is retained in species that live beyond a depth of 2,000 meters.

Benthic fishes of the deep sea, like their counterparts
of shelf waters, have no swimbladder and are in the habit

of resting on the bottom. The most diverse forms are chlorophthalmids, bathypteroids, zoarcids, liparids, and ogcocephalids. Curiously enough, common benthic fishes of shelf waters, such as scorpaenids, cottids, blennioids, toadfishes, and cling-fishes, have few or no representatives in the deep sea. But sea snails (Liparidae) and eelpouts (Zoarcidae) have invaded deep waters, mostly in temperate and polar regions. In Antarctic waters there is also a deep-water family (Bathydraconidae) of notctheniiform fishes. From warm temperate to tropical parts of the deep ocean most benthic fishes belong to the families Chlorophthalmidae (on the slopes only), Bathypteroidae, Ipnopidae, and Ogcocephalidae. But eleven species of zoarcids have been taken at depths from 830 to 3,280 meters in the Panamanian area (Garman 1899). During periods of up-welling this area has a high primary productivity; so, during the summer months, have the subpolar and temperate waters, where most deep-sea zoarcids live. Is their distribution related rather to their habits than to other factors? Like their relatives from coastal waters, deep-sea zoarcids may well get some of their prey by lying in wait for it, then suddenly seizing it.[20]

Such habits may not give a fish adequate living under the warmer surface waters of the ocean, where primary productivity is mostly quite low. For instance, in the deeper parts of the Pacific between 35°N and 50°S the dominant groups of benthic animals are "sponges, foraminifera, mostly agglutinated, nematodes, small sipunculids and polychaetes, sessile scyphoid polyps, small madreporarian corals, bivalve and less frequently, gastropod molluscs and various small crustaceans (Amphipoda, Isopoda, Tanaidacea, Cumacea). The background of this impoverished, small-sized and very thin bottom fauna is usually formed by the remains of animals inhabiting the overlying waters: empty siliceous tests of radiolarians, squid beaks, shark teeth and fragments of the oozes" (Filatova and Zenkevitch, 1966). To collect enough food in such surroundings requires much exploration.[21] If so, it is not surprising that bathypteroids have well-developed tracts of red fibre in their myotomes.

This discussion of the structure, habit, and habitats of

20. Species of *Lycodes* feed on benthic crustaceans, polychaets, bivalves, echinoderms, and fishes (Andriashev, 1964).

21. Bathypteroids are known to feed on polychaets, copepods, cumaceans, and amphipods.

benthic deep-sea fishes must, of course, be tentative in view of our limited knowledge. At all events, if the success of a group is measured by its overall living space and the numbers of its species and individuals, then the most successful bottom dwellers in the deep sea are macrourids, brotulids, and morids. And these fishes work, rather than wait for, their living.

CONCLUDING CONSIDERATIONS

These studies of deep-sea fishes remind us of Rashevsky's concept of optimal design: the hypothesis that "biological structures which are optimal in the context of natural selection, are also optimal in the sense that they minimize some cost functional derived from the engineering characteristics of the situation" (Rosen, 1967)[22]. In considering their organization, we have largely been concerned with the "engineering characteristics" of deep-sea fishes in relation to their physical and biological environments. In the context of natural selection optimal design is likely to be a balance between opposing stresses. Mayr (1963) puts it thus: "Since the genotype as a whole is an inter-acting, integrated system, virtually all aspects of the phenotype are the result of a compromise between opposing selection pressures."

Consider a single population of a deep-sea fish. The main selection pressures acting on this population are subsumed by these essentials: (1) living space must be adequate; (2) enough individuals must avoid being eaten; (3) individuals of the population must keep in touch; and (4) reproduction must maintain the population above some minimal size and density. These four desiderata are, of course, interrelated. But until we can define the populations of deep-sea fishes, which may well be extensive, little can be said of their living spaces (Marshall, 1954). Under the second heading we have already considered their use of cover, camouflage, and ventral light organs. We ought to know more of the relative development of the Mauthnerian apparatus, which

---

22. Rosen equates the cost of an organ to $I + E$, where I, the intrinsic cost, is the energy needed to form and maintain the organ, and E, the extrinsic cost, is related to selection pressures [a cost "reflected in the overall fecundity of organisms bearing the organ in question, as compared with (ideally) other organisms differing from those given only in the structure of that organ"].

in fishes from other environments helps them to escape quickly.[23] Keeping in touch involves *inter alia* various forms of signalling and the apt elaboration of sense organs. With regard to communication between the sexes, we saw that (1) many mesopelagic fishes have sexually dimorphic patterns of light organs; (2) females of the dominant groups of bathypelagic fishes produce olfactory signals; and (3) males of the dominant groups of benthopelagic fishes are sound producers.

The reproductive adaptations of mesopelagic and bathypelagic fishes are correlated with their life history patterns. The eggs, which are probably shed and fertilized at depth, develop as they rise toward the surface. The larval existence is certainly passed in the euphotic zone, where the young find such suitable food as larval invertebrates and small copepods, dependent themselves on phytoplankton. During and after metamorphosis the young move down to the adult living space. The eggs and young of bathypelagic fishes thus run a greater vertical gauntlet of physical changes and predation than those of mesopelagic species. A population of a black *Cyclothone*, a gulper eel, or of an anglerfish must thus have an overall fecundity to more than offset relatively great inroads of mortality. But we have seen that the organization of bathypelagic fishes is pitched at a level to conform to their food-poor surroundings. How, then, do they manage to produce enough eggs? We should keep in mind that recent work on the California sardine indicates that reproduction accounts for only about 1 percent of the energy consumed during its life (Anon., 1968).

When fully mature, female anglerfishes range from about 60 to 600 mm in length and the ripe eggs are between 0.5 and 1 mm in diameter (Bertelsen, 1951). Thus, if a 60 mm *Cyclothone* produces about 10,000 eggs, probably nearer 0.5 mm than 1.0 mm in size, the fecundity of female ceratioids must be numbered in hundreds of thousands of eggs. But ceratioids do not have an extended spawning season. Catches of larvae in the North Atlantic indicate that they either spawn during the spring or summer months. Even so, female ceratioids, which

---

23. We might expect this system to be prominent in mesopelagic fishes, but to be reduced or regressed in benthic species. I was unable to find a Mauthnerian system in a female ceratioid (*Melanocetus*), but it is said to be present in *Cyclothone*. *Cyclothone* spp. are not only eaten by other bathypelagic fishes but also by oplophorid prawns.

have few enemies, probably live to a considerable age (Bertelsen, 1951).

In Japanese waters the spawning season of *Gonostoma gracile* lasts from September to April. Here Kawaguchi and Marumo (1967) made these remarkable discoveries. About a year after hatching the fish attain sexual maturity as males and they mate with females that are two years old or more. In the summer of their second year these males become females, which are more than 70 mm in standard length. This species thus makes the best possible use of its mature biomass. It would not be surprising if *Gonostoma bathyphilum* and *Cyclothone* spp. prove to have a similar life history. If so, the change from male to female will not only involve the reproductive complex but also a transition from macrosmatic to microsmatic forms.

In food-poor waters, it is best for a predator to take the widest possible size-range of prey species. Bathypelagic fishes are so adapted. *Cyclothone* spp., the smallest members of this fauna, feed on organisms ranging in size from copepods to small fishes (other *Cyclothone*). Large female ceratioids take prey ranging from copepods to squids and fishes. There are even records of their capacity to swallow fishes two or three times their own length (Bertelsen, 1951). But predators with a very restricted food supply must not be too large. The greater their size, the less the chances of sexual encounter. To a remarkable extent, the successful groups of bathypelagic fishes have "got around" this problem by evolving small, quickly maturing, short-lived males. In the most specialized forms the males even become long-lived by becoming parasitic on the female, as in certain, but not all ceratioid anglerfishes. In *Gonostoma gracile*, at least, the males extend their lives by becoming females. These fishes have surely made the shrewdest compromise with opposing selection pressures. Of the mesopelagic fishes other than the light-colored forms of *Cyclothone*, only the species of *Idiacanthus*, and they are voracious forms, have evolved dwarf males. The problem of sexual encounter is, of course, greatest in predatory species, each individual of which needs more living space than one belonging to a plankton-eating kind. Is this why ovotestes are developed in paralepidids and *Alepisaurus* (Mead, 1960)? If the sexes fail to meet, can these species turn to self-fertilization?

Benthopelagic deep-sea fishes, most of which are much larger than bathypelagic species, must be much more fecund. Moreover, the slope-dwelling species, living in relatively rich waters, are quite abundant. Because most species produce floating eggs, the over-riding selection pressures behind the life-history patterns of the slope-dwellers are related to adequate maintenance of their populations. The ocean is always on the move, and seen from above, the slopes are relatively narrow strips of the deep-sea floor. Larvae in the surface waters may soon be carried away from the slopes. The macrourids, at least, have countered such wastage by producing larvae that avoid the waters above the seasonal thermocline. Just below the thermocline the circulation is likely to be most sluggish and food relatively abundant (Marshall, 1965). And there is some evidence that their early life (until metamorphosis) is quickly passed. In these two ways, at least, the slope-dwelling macrourids have met the problem of the narrow dropping zone beneath their descending recruits. Clearly they have met the selection pressure very well, for macrourids, and other benthopelagic fishes, are most diverse over these zones, where food is quite abundant.

Mesopelagic and bathypelagic fishes, which live over much greater expanses of the ocean, have more freedom for larval maneuver.[24] Their early life is passed in the surface waters. Virtually nothing is known of the early life of benthic deep-sea fishes. The chlorophthalmids, which are slope dwellers and apparently produce floating eggs, must face the same problems as the macrourids. The zoarcids and liparids, like their shallow-water relations, presumably lay large, heavily yolked eggs, which produce large, advanced larvae. Nothing is known of the early life of bathypteroids and ipnopids. We have seen how bathypteroids are designed to live even over impoverished parts of the deep-sea floor. Both groups of fishes seem to have a patchy distribution. Indeed their populations may well consist of relatively few individuals. At all events, one may argue that the sexes do not always succeed in meeting, for both bathypteroids (Mead, 1960) and ipnopids (Nielsen, 1966) contain ovo-

24. For certain mesopelagic fishes this freedom may be such that a species will breed only in a particular region of its living space (O'Day and Nafpaktitis, 1967).

testes and are so potentially capable of self-fertilization.

Deep-sea fishes are thus even more adaptable than we once supposed. But in some respects their environments are less exacting than those of fishes in shallow temperate seas. Most kinds of deep-sea fishes reproduce between the subtropical convergences (between about 40°N and 40°S) where oceanic waters are particularly stable in these ways: (1) primary productivity, and hence the biomass of deep-sea life, does not fluctuate widely from season to season and probably not from one year to the next; (2) physical properties are relatively constant; and (3) ecological systems are stable through their considerable taxonomic diversity. Temperate waters are less stable in all three respects, which means that their biota must at least be adaptable enough to cope with fluctuating selection pressures.

4 Life and Pressure in the Deep Sea

After his dredging expedition to the Aegean Sea, Edward Forbes introduced his concept of an azoic zone to members of the British Association, who were meeting at Cork in 1843. His Greek work, combined with his experience in British waters, led Forbes to conclude that there was no life in the sea beyond a depth of some 550 meters. Despite evidence to the contrary (Marshall, 1954), Forbes' idea seemed reasonable. Wyville Thompson (1887) wrote, "it was almost as difficult to believe that creatures comparable with those of which we have experience in the upper world could live at the bottom of the deep sea, as they could live in a vacuum, or in the fire." But more and more evidence showed that the very depths of the ocean contained diverse forms of life. Indeed, it was not long after the Challenger Expedition (1872–76) that the voyages of two French research ships, the *Travailleur* and *Talisman*, led to the first tests concerning the effects of high hydrostatic pressures on aquatic life.

During the *Talisman* dredging expedition of 1882–83, organisms were taken that must have lived at depths of about 6,000 meters. This discovery stimulated Regnard and his colleague Certes to begin (in 1884) their series of pressure tests, which were brought together in Regnard's (1891) monograph. They devised a chamber with quartz windows, which could hold a pressure of 800 atmospheres for several hours. Small aquatic invertebrates, such as *Cyclops*, *Daphnia*, and *Gammarus*, became more active when subjected to changes in pressure to a level of 100 atmospheres. Above this pressure, they were tetanized, but if the period of compression was short, they recovered on decompression. At pressures of some 500 atmospheres, the activities of yeasts, algae, fish eggs, tadpoles, and muscle were depressed. But fish eggs developed normally after 6 hours exposure to a pressure of 200 atmospheres. Marine and freshwater fishes became immobile, or might be killed, by pressures as low as 300 atmospheres.

Life and Pressure in the Deep Sea

Perhaps the most interesting discovery was that the activity of organisms increased after modest rises in pressure. During this century technical advances have made possible the application of hydrostatic pressures as high as 100,000 atmospheres. Marine biologists are, of course, concerned largely with the effects of pressures ranging to some 1,000 atmospheres, the level exerted at the greatest depths. But it was not until the late 1920's and the early 1930's, that such problems were resumed. This research and subsequent endeavors are well reviewed by Cattell (1936) and Johnson, *et al.* (1954).

Molecular biologists may well be concerned with the effects of high hydrostatic pressures on the large molecules of living organisms. Changes in molecular volume are evidently involved, but how are these related to the biological changes observed? Biophysicists may, for instance, concentrate on changes of cytoplasmic viscosity or on the changed properties of muscles that are associated with changes in pressure. Here, though, our concern is with whole organisms.

GAS-CONTAINING ORGANISMS

If organisms contain gases, changes of pressure could have marked effects. Indeed, the sensitivity of marine invertebrates to small changes of pressure may well depend on the presence of gas layers or vesicles on outer surfaces, particularly those containing quinone-tanned proteins (Digby, 1967). Deep-sea animals producing relatively large reservoirs of gas belong to three major groups; siphonophores, cephalopods, and teleost fishes. The physonect siphonophores are headed by a buoyant, gas-filled float, but some or even most of their buoyancy, resides in their gelatinous bracts, which are lighter than sea water (Jacobs, 1937). Parts of physonects have been found in closing nets fished below 1,000 meters. Whole colonies have been seen from underwater vehicles at depths of the scattering layers off San Diego (Barham, 1966). In fact, acoustic records may be correlated, at least in part, with the presence of physonect siphonophores.

In the San Diego region, *Nanomia bijuga* appears to be the dominant physonect, and its production of gas and bubbles has recently been studied by Pickwell (1967). Observations from the deep submersible DR/V Deepstar showed that the pneumatophore of each *Nanomia* is inflated at all times. The gas in the float, which is almost pure carbon monoxide, is presumed to come from an enzyme system in the cells of the gas gland. In floats with a high rate of CO production, 3 or 4 molecules of oxygen are equivalent to each molecule of CO secreted. Some of the energy of oxidation is presumably needed to keep the pneumatophore inflated at the ambient pressure of 30 to 45 atmospheres. When the colony moves toward the surface—some were seen at depths between 50 and 100 meters—gas must be released through the apical pore of the float.

The release of bubbles by *Nanomia* reminds us of analogous behavior in teleosts with a pneumatic duct. But all deep-sea fishes with a swimbladder have a closed sac. At mesopelagic levels (about 150–1,000 meters), species with a swimbladder belong to these main groups; stomiatoids, argentinoids, myctophids, melamphaids, chiasmodontids, and trichiuroids. Except for the black species of *Cyclothone*, which have a gas-filled swimbladder when young, but which gradually regresses and is invested with fat as they mature, bathypelagic fishes (about 1,000–3,000 meters) have lost the swimbladder entirely. Most members of the benthopelagic fauna, notably halosaurs, notacanths, synaphobranchid eels, macrourids, morids, and brotulids, have a well-developed swimbladder. But this organ is absent in benthic deep-sea fishes, just as it is in their counterparts of coastal seas. (See also Marshall, 1960, 1965a.)

The loss of the swimbladder in bathypelagic fishes is hardly related to high ambient pressures—for benthopelagic fishes living at depths down to 7,000 meters retain this organ—but is rather the focal point of their much reduced organization to conform to a food-poor environment. But if we consider mesopelagic and benthopelagic fishes together, there is one feature of their organization that is directly related to their depth ranges, and hence, presumably, to the pressures around them. This is the length of the capillaries in their retia mirabilia, which circulate blood to the gas gland of the swimbladder. In upper mesopelagic species (about 150–500 meters) the retial length ranges from 1.0–2.0 mm. Lower mesopelagic species and those from comparable benthopelagic levels (about 500–1,000 meters) have retia from 2.5 to 6.0 mm in length. Benthopelagic species ranging from

Life and Pressure in the Deep Sea

depths of about 750 to 1,500 meters contain retia of 10–15 mm, which length is increased to about 25 mm in species from depths of about 1,500–3,500 meters. (See also Marshall 1960, 1965a.)

Two main lines of research show that it is reasonable to link retial length to ambient hydrostatic pressures. Scholander and his colleagues (Scholander, 1958) have demonstrated very well that one function of the retia is to maintain the difference between gas tensions in the swimbladder and those in the blood entering the retia (about 1 atmosphere). This blood, as it flows along hundreds of arterial capillaries, is separated by capillary walls from the outflowing blood in contiguous venous capillaries. When the outflowing blood enters the retia capillaries it holds gases (largely oxygen) at tensions corresponding to those in the swimbladder. Thus, during the passage of blood along the venous capillaries gases will diffuse into the arterial blood. Moreover, this process is very efficient. Clearly, the greater the depth the steeper will be the gradient of gas tensions and the more the need to maintain the efficiency of the counter-current exchange of gases. The latter depends *inter alia* on the rate of diffusion across the two sets of capillaries and indirectly on the rate of blood flow. Increase in length of the capillaries will increase the surface area for diffusion and also slow down the flow of blood; hence, the direct correlation between the depth ranges of fishes and the lengths of their retial capillaries.

The other line of research, due to Kuhn and his colleagues (1963), has revealed how a rete mirabile can build gas tensions through counter-current multiplication. "The gas gland is considered to introduce a chemical, most likely lactic acid, into the blood. This chemical will pass out in the blood of the venous capillaries of the rete, and in doing so, will decrease the solubility of the blood gases. The released gases will diffuse into the blood of the ingoing capillaries and gradually accumulate at the gas-gland end of the rete. And, as the blood continues to circulate, the pressure will steadily increase. By this salting-out process any kind of gas will be released, but oxygen will also be produced by the Root and Bohr Effects."

"Considering the salting-out process alone, the multiplying power of a rete will directly depend on its length and the degree of permeability between the venous and arterial capillaries. The inverse factors are the speed of flow in the capillaries, the bore of the capillaries and the solubility coefficients of the gases" (Marshall, 1965a). Again, a counter-current process, this time of multiplication, is increased by lengthening the retial capillaries. Kuhn and his colleagues calculate that a rete 20 mm in length should easily build the gas tensions that must exist in the swimbladder of the deepest living fishes. This length is close to 25 mm, the maximum measured length of the retia of abyssal macrourids and brotulids.

To maintain and produce the requisite tensions of swimbladder gases, deep-sea fishes have thus evolved counter-current systems of a length to "match" the ambient hydrostatic pressures of their living spaces. *Spirula retroversa*, the only deep-sea cephalopod with a gas-filled float, maintains its buoyancy by very different means. When freshly caught, *Spirula* is close to neutral buoyancy (Denton, Gilpin-Brown, and Howarth, 1967). Except for the last chamber which contains fluid, the shell is filled with air at somewhat less than atmospheric pressure. Tests on the shell proved that it will withstand hydrostatic pressures up to some 50 atmospheres, which strongly suggests that *Spirula* lives down to depths of about 500 meters, not, as used to be thought, down to 1,000 meters or more. As the animal grows, new chambers are added to the shell. In fact, *Spirula*, like its relatives *Nautilus* and *Sepia*, develops an air-filled shell with chambers strong enough to withstand the ambient pressures of its living space. As each new chamber is added to the spiral shell, fluid is withdrawn (by an osmotic mechanism) when the walls have acquired the necessary strength. The timing of this change from fluid to air is presumably fairly critical. Denton and his colleagues point out that the earlier work on *Sepia* indicated that an osmotic pump is capable of withdrawing fluid against pressures exerted by the water column down to a depth of 240 meters. If, as seems likely, this must also apply to *Spirula*, how does it fare at depths below this level?

HYDROSTATIC PRESSURE AND THE VERTICAL DISTRIBUTION OF DEEP-SEA LIFE

Bassogigas profundissimus, a brotulid fish with a well-developed swimbladder, has been taken at a depth of

Life and Pressure in the Deep Sea

7,160 meters (Nielsen and Munk, 1964). At this level the gases in the swimbladder must have a specific gravity of about 0.7 and thus still give appreciable lift. Beyond 8,000 meters a swimbladder will have little or no buoyancy, though there are no certain records of fishes, with or without a swimbladder, much beyond 7,000 meters. But various foraminiferans, actinarians, polychaets, nematodes, echiuroids, prosobranch gastropods, pelecypods, isopods, amphipods, and holothurians have been taken at a depth of 10,000 meters or more. Pogonophorans and crinoids almost reach this level, but the deepest records of echinoids, asteroids, and ophiuroids are between 7,000 and 8,000 meters. Pycnogonids, cirripedes, and decapods (Anomura) are not known beyond 6,860, 6,840, and 4,340 meters respectively.

The deepest reaches of the ocean flow thus support a rather limited fauna. Indeed, assemblages of species at depths below 6,000 meters have been judged to be distinct enough to deserve a faunal name: hadal or ultra-abyssal. Belyaev (1966) states that some 300 species of hadal animals have been described, and he estimates that the number may well rise to about 700 after specialists have studied the relevant catches. More than 1 in 2 of these species are endemic, though Menzies and George (1967) have shown how difficult it is to define a separate hadal fauna. This problem need not concern us here, but there is the question of the possible effect of high hydrostatic pressures on vertical distribution. Belyaev (1966) believes that "the tremendous hydrostatic pressure is a limiting factor restricting the diversity of the ultraabyssal fauna, while intensive sedimentation creates favourable feeding conditions in the deep-sea trenches; this results in a mass development of a number of species which have become adapted to life under conditions of high pressure." It should also be remembered that trench faunas consist largely of deposit-feeding forms. Thus, the absence of such animals as the decapod crustaceans and the asteroids may reflect their carnivorous proclivities rather than their physiological inability to cope with high hydrostatic pressures. In brief, the essential requirements for most forms of trench life are suitable means for living on deposits at pressures of 600 atmospheres or more.

The trenches contain numerous endemic species. These forms are thus stenobathic, but are they also stenobaric? There is no present answer, but we do know that ooze bacteria from great depths are not necessarily able to live just as well at atmospheric pressure. Indeed, barophilic bacteria, as they are called, grow properly only at high hydrostatic pressures (Zobell and Morita, 1956). Perhaps the same is true of the more complex forms of trench life.

Eurybathic forms are, of course, as eubaric as the known limits of their depth ranges. Some species, as Ekman's (1953) list shows, extend from littoral regions to depths of several thousand meters. For instance, the sea urchin, *Echinocardium australe*, and two brittle stars, *Ophiacantha bidentata* and *Ophiocten sericeum*, tolerate a range of over 400 atmospheres in hydrostatic pressure. The first species must also tolerate a temperature range of some 20°C.

Pressure tests on eurybaric forms, which ought to be quite viable, should be undertaken over a range of temperatures. In 1934, Brown (quoted by Johnson, *et al.*, 1954) showed very well the marked influence of temperature when he studied the effects of pressure on pectoral fin muscles of the red grouper (*Epinephelus morio*). The degree of contraction following increases of pressure was decreased at temperatures below 14–16°C, but increased at temperatures above this range. At 14–16°C, pressure increases up to 204 atmospheres had no effect on the size of contraction. Evidently, the preferred environmental temperature of the red grouper used in the experiments was close to 14–16°C.[1] The conclusion was that the influence of temperature on the effects of pressure was related essentially to environmental temperatures. Experiments by Brown, Johnson, and Marsland (1936) on *Photobacterium phosphoreum* and *Achromobacter* spp. may be interpreted in a similar way (Johnson et al., 1954). Above the temperature optima of these bacteria, pressure increases the intensity of their luminescence, but below the optima emission decreases. At the optima, pressure changes have little or no effect on luminescent levels. More recently, Schlieper (1968), who followed the effects of pressure changes on the oxygen consumption, ciliary movement, and enzyme activity of tissues taken from the mussel *Mytilus* and the shrimp *Crangon*, concluded that cellular resistance to pressure is highest at the optimum temperature of the species.

1. *Epinephelus morio* is the common grouper in West Indian and Florida waters, where, of course, temperatures are usually above 14–16°C.

Life and Pressure in the Deep Sea

Are there any natural "experiments" in the sea that might have a bearing on this aspect of pressure-temperature relations? A number of widely distributed deep-sea animals certainly live over wide ranges of pressure, combined with narrow limits of temperature. We might think first of submergence, whereby animals with a wide longitudinal distribution have a definite tendency to live below the warmer waters that cover part of their range. One instance is the brittle star, *Ophiacantha bidentata*, which in high arctic waters occurs up to 5 meters; in low arctic regions to 23–30 meters; in boreal regions to 200 meters; and in the central Atlantic at abyssal levels. Ekman (1953), who gives this instance, considers abyssal regions as those beyond a depth of 1,000 meters. Even more striking is the holothurian, *Elpidia glacialis*, which ranges from 70 meters in coastal waters of the arctic to over 7,000 meters in the deep sea. Thus, it would seem seem that these two echinoderms can tolerate wide ranges of pressure at ambient temperatures not exceeding about 5°C.

For other instances of submergence in benthic invertebrates, Ekman (1955) should be consulted. Mid-water forms also show submergence, a classic example being the arrow-worm, *Eukrohnia hamata*. In the Southern Ocean it is commonest between the surface and 200 meters but does extend down to about 1,500 meters. In equatorial regions it occurs largely below a level of 500 meters. Again, we are dealing with a cold-loving (psychrophilic) species that is at home over a relatively wide range of pressures.

This is also true if we consider the expatriated individuals of certain deep-sea fishes. For instance, in the North Atlantic, adult females of the ceratioid anglerfishes *Himantolophus groenlandicus*, *Ceratias holboelli*, and *Oneirodes eschrichti* are known as far north as 65° latitude in West Greenland waters. But the fertile distribution of ceratioids, shown by the distribution of their larvae, is below the warmer waters between about 40°N and 35°S (Bertelsen, 1951). In West Greenland and Icelandic waters the females of these species may reach a great size, but they do not reach sexual maturity: they are simply able to lead a vigorous vegetative existence during the productive season. Off Iceland, trawlers have taken the first and second species at depths less than 100 meters, which may be contrasted with their vertical distribution over the fertile parts of their range, where most individuals live between depths of 1,000 and 2,000 meters.

Once more, here are psychrophilic species that are also eurybaric. Where strong up-welling occurs, the same tendency may also be seen. In the Gulf of Guinea bathypelagic species follow the sharp upward inclination of the isotherms, so taking advantage of rich feeding grounds relatively close to the surface (Voss, 1967).

Lastly, warmth-loving (thermophilic) species may also be eurybaric. In the Red Sea, due to the shallow sill in the Straits of Bab el Mandeb, the waters are isothermal (near 21°C) at depths below 300 meters. Owing, it seems, to the unusual warmth of the deeper waters, certain decapod crustaceans, which are confined to shelf waters in the Indian Ocean, have been able to colonize the deep-sea floor of the Red Sea (Ekman, 1953). The shark *Eugaleus omanensis*, first taken in the Gulf of Oman at about 200 meters, is probably the same species as that photographed down to about 2,000 meters in the Red Sea (Marshall and Bourne, 1964).[2]

Diverse marine organisms, from bacteria to fishes, can thus tolerate wide range of hydrostatic pressure so long as ambient temperatures are at, or near, their optimal requirements. Eurybaric-thermophilic species are doubtless rarer than eurybaric-psychrophilic forms, but this may simply reflect the temperature structure of the ocean, which gives much more scope for the second kind of physiological association. It ought to be feasible, then, to keep some deep-sea animals in the laboratory, so long as they are kept at temperatures near that of their living space. In fact, Baker (1963) has shown this very well. For instance, in tanks chilled to 7–10°C, *Spirula* lived for 8 days, *Gigantocypris mulleri* for 13 days, *Acanthephyra* sp. for 18 days, and *Oplophorus* sp. for 31 days. The animals, which came from hauls with an Isaacs-Kidd Midwater Trawl (IKMT), were found to be in better condition when the haul was short (one half to three quarters of an hour instead of two hours) and taken at a slower towing speed (one and a half to two knots instead of two and a half to three knots). Moreover, Foxton (1964) obtained two ovigerous individuals of *Acanthephyra purpurea* from an IKMT fished to a

---

2. Stewart Springer has kindly drawn my attention to this.

Life and Pressure in the Deep Sea

maximum depth of 400 meters and he kept them in a two and a half liter tank at a temperature of 7–8°C. Not only did these deep-sea prawns survive for 20 days, but larvae also hatched from the eggs.

These tests are surely encouraging. Clearly, the first problem is to catch viable individuals, and here a thermally insulated bucket on the net should help. The insulation should be such that temperatures in the bucket should remain fairly stable as the net is hauled upwards through the seasonal thermocline. For mid-water animals that do not migrate into the surface mixed layer, thermal shock is likely to be most severe as they pass through the thermocline. The net should be towed, as we saw, at slow speeds and for relatively short periods, which presumably helps reduce damage to the animals in the net. Some kind of buffering material inside the net might also help.

Certain kinds of benthic animals are both eurybaric and eurythermic. This is, of course, true of mesopelagic animals that undertake vertical migrations. Moreover, such forms are most diverse in subtropical and tropical regions. Thus, a lantern-fish, a squid, or a euphausiid, after moving upward from a level where the pressure is, say 50 atmospheres and the temperature about 10°C, will, in a hour or so, when it has reached the surface waters, have experienced a change in pressure and temperature of nearly 50 atmospheres and about 10°C. Ontogenetic migrations of bathypelagic animals, though taking a much longer time, involve even greater changes of pressure and temperature. For instance, during and after metamorphosis, a ceratioid angler-fish in migrating down to the adult living space may have to tolerate a pressure increase of more than 100 atmospheres and a fall of temperature of some 15°C. Reverse changes of this order must be sustained during the ontogenetic ascent of the eggs and larvae to the surface waters.

Concerning one mesopelagic form, Teal and Carey (1967) have studied the effects of changing pressure and temperature on the respiration of a euphausiid, *Thysanopoda monacantha*. By day it seems to be centered at levels below 500 meters, and its upward migrations do not extend much beyond a level of 150 meters. Its responses were compared with those of epipelagic species (for example, *Euphausia americana* and *Meganyctiphanes norvegica*), which are concentrated above 500 meters by

97

day and migrate to the surface at night. Respiration of both mesopelagic and epipelagic forms was measured between 5 and 25°C and at pressures between 0 and 100 atmospheres. In most instances, pressure changes only influenced the rate of respiration of the epipelagic species at the higher temperatures (20–25°C), which, as Teal and Carey remark, are combinations of physical factors not found in the ocean. Concerning *Thysanopoda monacantha*, pressure changes increased respiration more at the lower temperatures than at higher levels. Teal and Carey conclude that the respiration of epipelagic euphausiids is determined by temperature alone and the rate decreases as they descend into cooler waters. The respiration of the mesopelagic species remains constant throughout its depth range. They compare their findings with those of Napora (1964) on the decapod, *Systellaspis debilis*. On increasing the pressure the metabolism of this prawn increased just enough to counter the depressant effects of decreasing temperature. Thus, as these animals move up and down in the ocean, their metabolism should tend to stay at a constant level.

When mesopelagic fishes can be kept in captivity, will comparable experiments yield much the same results? Observations from deep submersibles suggest that these fishes are inactive by day. For instance, several observers have been impressed by the quiescent suspensions of lantern-fishes (Backus, 1968). Most likely, these and other mesopelagic fishes are most active just before and after sunset, when they migrate and are seeking their food. Though such observations are no means of predicting the likely effects of changing temperature and pressure, they should help in the planning of experiments. In particular, care should be taken to keep the animals in dim surroundings, close to those of their natural environment.

Lastly, we do not yet know the sensitivity of any mesopelagic form to pressure changes. The threshold may be quite low for certain organisms from coastal seas. Thus, certain arthropods (*Nymphon* and *Corophium*) can perceive pressure changes of not more than 2.5 millibars per minute. For certain copepods, barnacle larvae, cumaceans, decapod larvae, and amphipods, the threshold is about 10 millibars (Knight-Jones and Morgan, 1966). In diverse planktonic forms, increased pressure leads to upward movements, and a decrease induces either active or passive descent. Such reactions must help these organisms stay at a particular level in the water column, and the same may be true of mesopelagic animals of the open ocean. Deep scattering layers—and mesopelagic fishes with a swimbladder are prominent scatterers—may change their daytime level after sudden changes in light intensity, as when clouds cover the sun. One layer observed by Blaxter and Currie (1967) rose about 100 meters by the end of a period of cloud cover lasting about 20 minutes. This and other evidence suggests that the animals of a particular scattering layer tend to stay at a constant level of light intensity, but this is not to say that these animals are insensitive to change of pressure. Relevant tests should be undertaken before very long.

5 Aspects of Convergent Evolution

"Nature diversifies and imitates" —Pascal, *Pensées*

A New English Dictionary on Historial Principles (Supplement, 1933) traces the biological concept of evolutionary convergence to the fourth edition of the *Origin of Species* (1866).[1] Here Darwin considers his correspondence with "a distinguished botanist," Mr. H. C. Watson, who "believes that I have over rated the divergence of character . . . and that convergence, as it might be called, has likewise played a part. This is an intricate subject which need not here be discussed. I will only say that if two species of two closely allied genera produced a number of new and divergent species, I can believe that these new forms might sometimes approach each other so closely that they would for convenience sake be classed in the same new genus, and thus two genera would converge into one" (Darwin, 1861). But this is not what later biologists have come to mean by convergence—the independent acquisition of similar characters in lineages that have evolved in the face of similar environmental conditions. The key word is "independent," meaning that the similar characters do not depend on any (probable) common ancestry. Darwin's sense of convergence is closer to our present idea of evolutionary parallelism, which Simpson (1961) defines as "the development of similar characters separately in two or more lineages of common ancestry and on the basis of, or channeled by, characteristics of that ancestry."

But parallelisms, which may well be commoner than we realize, are not always distinguishable from convergences. For instance, the angel-fishes and butterfly-fishes seem to be more closely related to each other than to any other group of percoid teleosts, and they are classified together in the family Chaetodontidae. Even so, Freihofer (1963), who found distinct patterns of the ramus

---

1. Darwin first considered "convergence" in the third edition of the *Origin of Species* (1861).

lateralis accessorius nerve complex in these two groups, suggests that "they may be excellent examples of convergence among coral reef fishes." On the other hand—and in spite of their many common features, including a *Tholichthys* stage in the early life history—some of the resemblances, particularly those that mark them so strikingly as "coral fishes," might be due to parallel evolution. Characters that were not present in their ancestry may have been acquired as a result of common genetic potentialities and selection pressures in coral-reef environments. But we shall be dealing here almost entirely with forms so remotely related that it is impossible or unlikely that their common ancestors could have endowed them with the congruences under discussion. The congruences will thus be convergences.

Darwin was well aware of the above sense of convergence but he used Owen's term, "analogy," which is similarity of overall function that is not dependent on common ancestry. Yet there are no hard and fast distinctions between convergence and analogy. "Convergent characters are analogous insofar as the similarity can be related to function, which is usually and perhaps always the case" (Simpson, 1961, p. 79).

A classic instance of convergence is the similarity in body and fin forms of cetaceans, ichthyosaurs, and fishes. Even more remarkable, perhaps, are the convergences between the eyes of cephalopods and vertebrates. Darwin was intrigued and troubled by the presence of electric organs in very different kinds of fishes. But the convergences between organisms are not only those of form and functional morphology. Species have biophysical, biochemical, physiological, and behavioral characters. Convergences may also arise at these levels.

There are manifold instances of convergence in plants and animals. Unicellular organisms may even mimic the organization of complex multicellular forms. *Acetabularia*, an alga with a single large nucleus, has "roots" and a "stem" that bears a spreading photosynthetic surface in the form of a cap. Certain convergences between plants are so detailed that Wardlaw (1965) calls them homologies of organization. Certain of the more complex ciliated protozoans, such as *Epidinium*, have parts corresponding in function to a mouth, oesophagus, rectum, anus, muscle fibres, nerve fibres, brain, and skeleton. These ciliates remind us most of certain rotifers. Organisms thus evolve convergences that transcend our major groups of classification. (The convergences between fishes, cephalopods, and malacostracan crustaceans are so striking that they will be considered fully at a later point.) The interstitial fauna of marine sands contains representatives of nearly all groups of invertebrates[2] which have converged in all or some of these respects: regressive evolution of body size; elongated shape; strongly contractile body wall, sometimes strengthened by a spicular kind of skeleton; adhesive organs (for example, epidermal glands), and prominent static organs (Swedmark, 1964). Orton's (1914) studies of the ciliary feeding devices on the gills of diverse gastropods, most lamellibranchs, *Amphioxus*, ascidians, brachiopods, and cryptocephalous polychaets showed them to be essentially similar, notably in the positions on the gills of the main food collecting cilia and of the lateral cilia that produce the main food and respiratory currents. Indeed, as Pantin (1951) remarked, some points of these convergences are so detailed that they almost mimic evolutionary homologies.

To cover properly the convergences within the major groups of animals is not the present concern. But further discussion of selected instances will form a background relevant to later treatment of convergence in fishes. Concerning arthropod evolution, Tiegs and Manton (1958) have emphasized the convergent aspects. A tracheal system evolved separately at least twice (Myriapoda, Insecta, and Arachnida), and an excretory Malphigian complex, though derived from different parts of the gut, is found in insects and arachnids. Mandibles and even a jointed exoskeleton are most likely to have evolved more than once. Despite the detailed congruences between the compound eyes of crustaceans and insects (multiple ommatidia, each with a lens, four vitreous cells producing a cone, and, usually, 7 retinular cells deployed around a central rhabdome), it is most probable that these eyes were evolved separately in each group.[3] One of Tiegs and Manton's arguments is that "we have either to admit a

2. Protozoa (foraminiferans and ciliates), Cnidaria (mainly hydrozoans), Turbellaria, Rotifera, Gastrotricha, Nematoda, Archiannelida, Polychaeta, Tardigrada, Crustacea, Mollusca (particularly opisthobranchs), Bryozoa, Echinodermata, Ascidiacea (synaptids).

3. This resemblance is reminiscent of the close similarity in fine structure of cilia belonging to diverse plants and animals.

detailed convergence or to avow an intimacy of relation between Crustacea and higher insects, such as is plainly inconsistent with the many profound differences between them."

Convergences in the molluscs are legion. In the gastropods there are diverse adaptive types, such as "limpets," "slugs," and terrestrial snails, which far transcend the bounds of classification. Indeed, about the very diverse mesogastropod division of prosobranchs, Morton (1958) wrote, "Every adaptation we shall later find in the opisthobranchs and pulmonates, the mesogastropods seem somewhere to have attempted for themselves." Thorson's (1946) three larval types have evolved independently in the major groups of molluscs. Long-lived planktotrophic larvae, which have an elaborate collecting and sorting system of food-gathering cilia, are found in most lamellibranchs and many prosobranchs. Short-lived planktotrophic larvae, bearing a much simpler ciliary system, are produced by many nudibranchs and various prosobranchs. Larvae depending on a large supply of yolk (lecithotrophic type) are known in the Amphineura, Scaphopoda, protobranchiate bivalves, and some gastropods.

Concerning other larval convergences, there are striking similarities between frog tadpoles of distantly related families (Orton, 1953). For instance, tadpoles that live in swift streams have a large, splayed snout, vacuum-cup mouthparts, and well-developed axial muscles (for example, certain species of Ascaphidae, Leptodactylidae, Bufonidae, Hylidae, and Ranidae). Tadpoles adapted for feeding at the surface develop upturned, papillae-fringed mouth-parts (for example, certain Microhylidae, Pelobatidae, Hylidae, and Dendrobatidae).

To conclude these background studies, we need do no more than recall two classic series of convergences in mammals. The first is the evolution of certain adaptive types (wolves, moles, flying-squirrels, anteaters, and so on) in placentals and marsupials. Less well known, but just as striking, are the convergences between the mammals of North and South America (for example, shrew-like, wolf-like, camel-like, and horse-like types). For further details, Simpson (1949) may be consulted.

A relevant transition to fishes may be made by considering the convergences between them and two invertebrate groups, the cephalopods and malacostracan

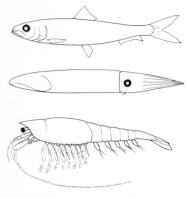

35   Convergence of form in a sardine (*Sardinops*), a squid (*Dosidicus*), and a decapod crustacean.

crustaceans. These convergences involve, *inter alia*, their shapes, muscular and nervous systems, sense organs, and photophores. With regard to shape, actively swimming species have something of a fusiform, streamlined appearance (Figure 35). To travel quickly through water, an inert medium eight hundred times denser than air, such a shape is needed. Strong, quick muscles are also needed, and one provision has been the (independent) evolution of striated muscle fibres in these three groups. Striated muscle of a goldfish has a contraction time of 35–160 m sec., which may be compared with 60 m sec. for squid mantle muscle and 40 m sec. for the abdominal flexor muscle of the crayfish (*Astacus*).

The most powerful main motor muscles of actively swimming fishes, squids, and diverse malacostracan Crustacea[4] are innervated in part by a giant fibre system. In fishes there is a pair of giant neurons (Mauthner cells) near the middle of the medulla oblongata at the level of the auditory nerve roots. The two giant axons decussate near their origin and then descend in the medial longitudinal fasiculus to the end of the spinal cord. Branches of these axons make synaptic contact with primary motor neurons at all levels of the cord (Figure 5). Squids have a pair of giant cells at the posterior end of the pedal ganglion. Their axons fuse and cross and then make contact in the visceral part of the brain with cell processes running to the mantle nerves. Each mantle nerve serves a stellate ganglion, from which nerves, each containing one giant fibre, pass to the muscles of the mantle. The kind of giant fibre system in the Malacostraca may be illustrated by that of the crayfish, which has three kinds of fibres: median, lateral, and motor. The paired median fibres come from the brain and after decussation there pass to the ventral nerve cord. The lateral fibres are in thoracic and abdominal ganglia, and like the median fibres, send branches to motor-giant fibres in each segment. The motor fibres, which have their cell bodies in one segment, decussate and then pass out in the posterior nerve of that segment.

Stimulation of the giant fibre system elicits these reactions. In fishes the tail is suddenly swept from side to side. There is a quick and overall contraction of the mantle muscle in squids, so water is forced from the

4. Giant fibres are also found in some ostracods and copepods.

Aspects of Convergent Evolution

mantle cavity through the funnel. In malacostracans the abdomen is suddenly flexed and the antennae are drawn together.

Giant fibres transmit impulses quicker than do smaller kinds of nerve fibres. For instance, giant fibres of the catfish, *Ictalurus nebulosus*, ranging from 23 to 43 μ in diameter, conduct impulses at 50 to 60 m/sec. The squid *Loligo*, having fibres of 280 to 400 μ conducts at 20m/sec., and in the crayfish *Cambarus* median fibres of diameter 100 to 250 μ conduct at 15 to 20 m/sec. Moreover, the neural links from sensory to motor elements are very short, and because of their size, the giant fibres innervate the entire mass of muscle. There is thus a quick and powerful motor response impelling a rapid escape reaction. The fish darts away; the squid shoots backwards or forwards, depending on the angle of the funnel; the crustacean flips backwards. By these convergent neuromuscular means enemies may be eluded, and the escape movements must quickly and effectively counter the inertia of water. Hence the need for a full and powerful, as well as a rapid, use of locomotory mechanisms. Teleosts that lead concealed lives on the bottom have a reduced Mauthnerian system. Octopods do not have a giant fiber system, nor do branchyuran crabs— and they too are bottom dwellers.

As for their brains, there are certain striking convergences between fishes and cephalopods. Here only one will be noted, concerning the glial cells, which may be involved in learning. Young (1964) writes: "It is perhaps relevant to notice that there is a really remarkable similarity in the glia found in nervous systems as distinct as those of cephalopods and vertebrates, although these animals have evolved independently for 500 million years or more. The space between the neurons of an octopus is packed with glial protoplasmic masses strikingly like those of vertebrate astrocytes. Evidently the operations of the nervous system depend essentially on such tissue."

Convergence between the eyes of vertebrates and cephalopods is well known. There is an essential similarity in the bauplan of the dioptric parts and the visual cells. In the former, fishes strikingly resemble cephalopods. Since water has a high refractive index, the cornea is not available as a major refracting surface. Instead, the fish and cephalopod eye refracts light through a spherical crystalline lens (Denton, 1960). There is also a fine congruence in the pattern of visual cells. In teleosts the cones and rods are arranged in two mosaic types, rows and squares (Engström, 1963). The basic similarity to cephalopods, as seen in the eyes of *Octopus*, is that the visual cells are disposed along criss-crossing lines, the two directions of which correspond to the vertical and horizontal axes of the animal. Such a pattern seems to be particularly well fitted for the detection of movements in the outside world.

Lastly, there are certain essential similarities between the statocysts of cephalopods and the ears of fishes. The elaborate cristae in the former, like the semicircular canals of fishes and other vertebrates, cover three planes.[5] The supposition that the cristae also detect angular accelerations in different directions was substantiated by Dijkgraaf, who showed that octopuses are able to do this, but not if the statocyst is damaged. There is also a sensory macula in the statocyst, which bears a statolith. The correspondence with the macula-otolith system of vertebrates is obvious. Evidently, the cephalopod macula is also a static dynamic receptor, providing in particular, a continual check on the position of the head and eyes (Young, 1964). In general, then, we can surely agree with Pantin (1951): "Nothing is more remarkable than the essential similarity of the 'behaviour machines' which have been independently evolved in the brains and sense organs of the most intelligent animals."

Evolutionary convergences are thus elicited by common ecological conditions. These conditions may simply be related to existence in a particular medium. A fusiform body, certain fin features and quick escape mechanisms are correlated with efficient motion in water—a heavy, highly viscous fluid. The independent acquisition of tracheae by certain arthropods is related to how animals with a chitinous exoskeleton, largely impermeable to oxygen, are to breathe in air. The evolution of tubes, strengthened with cuticle and, of course, open to the air, seems an obvious, perhaps the only, solution to this problem. More restricted ecological conditions, such as

---

5. As von Békésy (1967) says, the vestibular organ of fishes is a natural application of Lagranges' principle of describing the rotation of the body, because any rotation of the fishes' head is analyzed by 3 semicircular canals with axes along 3 perpendicular coordinates. Cephalopods have also discovered Lagrangian analysis.

existence between grains of sand or in swift streams may produce strikingly similar convergences between organisms of remotely related groups. We have also seen that convergences arise at different levels in the organization of living organisms.

Are there convergences at the levels studied by biochemists and molecular biologists? Concerning the so-called informational molecules (DNA, RNA, and proteins), which determine biochemical reactions, ontogenetic events, and evolutionary events, Zuckerkandl and Pauling (1965) argue that convergences (pseudohomologies are unlikely. They conclude: "Whereas one cannot say that pseudohomologies between polypeptide chains never arise in evolution, it seems very unlikely that such a process has occurred in the case of any particular polypeptide chain one is considering. It is thus likely that proteins as different in amino-acid sequence as mammalian haemoglobin and myoglobin indeed derive from a common molecular ancestor, not because of the characters of the amino-acid sequence they have in common, but because of the common characters of primary structure, tertiary structure and function taken together. The ease with which variations of a given type of protein can be produced through duplication and mutation of a gene should be so much greater than the ease of convergent evolution from independent starting points that on the basis of this consideration above any variants within a given type of tertiary structure and function seem to have a much greater chance to be phyletically related than unrelated." On the other hand, Wald (1965) considers that some convergence at molecular levels is likely because "faced with well nigh universal problems organisms everywhere may tend to gravitate towards common solutions, types of molecule that within the bounds of organic structure may represent optimal or near optimal conditions." Simpson (1964) even suggests that convergences in single kinds of molecules are more likely than in phenotypic characters which involve, of course, the interaction of myriads of molecules of many types. Selection, which produces diversity, does not usually act on single genes, but rather on higher levels of organization. As Simpson argues, in moving away from the genes, the nearer we are likely to come to selective action. Even so, Neurath, Walsh, and Winter (1967) use the concepts of homology and analogy in their endeavor to trace the evolution of structure and function in the proteases. By "analogy" they mean similarity in function without regard to structure. They conclude that "one of the striking features of the proteolytic enzymes as a group is the immense variety of biological functions served by enzymes employing one of a few basic mechanisms. For example, in the higher animals, enzymes for the activation of zymogens (trypsin), for digestion of dietary proteins (trypsin, chymotrypsin, elastase), for blood clotting (thrombin), for cytolysis (plasmin), and for sensing pain (kallikrein), all appear to use the same mechanism of gene duplication and subsequent divergent evolution. Equally striking is the variety of chemical solutions of the same functional problem, such as the peptide-bond cleavage by sulfhydryl proteases on the one hand and serine proteases on the other."

Moreover, certain molecular convergences are related to molecular function *and* common ecological conditions. For instance, in two series of fishes from cold and tropical waters, the temperature for thermal shrinkage of their collagen rises and is directly correlated with the hydroxyproline content (Prosser, 1965). Distantly related deep-sea fishes from the twilight zone have independently acquired golden forms of rhodopsin in their visual cells. These golden pigments (chrysopsins) are most sensitive to blue-green light, such as penetrates to about 1,000 meters in the clearest oceanic waters. Both kinds of adaptation seem to involve relatively slight changes in protein structure, which are presumably determined by informational molecules.

But what are we to say of the close resemblance between the bile salts of *Latimeria*, *Protopterus aethiopicus*, and the cyprinid teleosts? Haslewood (1964) is inclined to relate their common possession of a bile alcohol (cyprinol) to their ancestral or present, freshwater habitats. Yet it is curious that *Latimeria* has retained a freshwater type of bile alcohol. At all events, biochemical congruences are not always correlated with ecological similarities. For instance, the insulin composition of sperm whales (which is quite different from that of sei whales) is identical with the composition of pig insulin.

Fishes (and other organisms) thus display *some* molecular convergences that are related to common ecological factors, and we turn now to convergences at other level

Aspects of Convergent Evolution

of their organization. Consideration of electrical fishes, one of Darwin's "difficulties", makes an excellent introduction. Electric organs have evolved independently in six groups of fishes: skates (Rajidae), torpedo-rays (Torpedinidae), mormyroids, gymnotoids, electric-catfish (*Malapterurus*), and stargazers (*Astroscopus*) (Figure 36). The individual units (electroplaques), which are generally flattened and wafer-like, are derived from striated muscle fibres. One face of an electroplaque is innervated; the opposing face is thrown into deep folds. At a finer level the membrane of this folded surface continues deeply into the body of the electroplaque as a reticulum of tubules. This reticulum, which is evidently an essential electrogenic device, has been found in the electroplaques of all groups of electric fishes. In most electric fishes: skates, torpedo rays, stargazers, and gymnotids, the innervated surface of each electroplaque receives an elaborate network of nerve endings. But in mormyrids, the electric catfish, and one gymnotid (*Hypopomus artedi*), the innervation is through one or more stalks that extend from the electroplaque (Figure 37).

These convergences at the organelle and cellular levels show what must be done to turn striated muscle fibers into electroplaques. The great elaboration of the un-innervated surface is essential, whereas there are two main ways of innervation. At the next higher level of organization, the packing of the electroplaques to form a powerful electric organ depends on the nature of the medium. In the torpedo-rays, the electroplaques, which are stacked in vertical columns, are connected in parallel. In this way, enough power to stun prey can be generated in a low resistance medium. The stargazers (*Astroscopus*) also have vertically packed electroplaques that are connected in parallel. But to generate high electric power in freshwaters, which have a relatively high resistance, the necessary voltage is obtained through series connection of the units, as in the electric eel.

At the levels of comparative embryology and anatomy, the electric organs of skates, mormyrids, gymnotids, and the electric catfish converge in that they are derived from

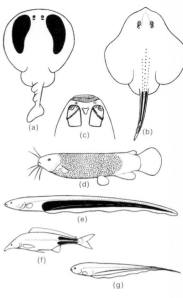

36 Electric fishes. (a) a torpedo-ray; (b) a skate (electric organs shown black); (c) upper part of head of a stargazer (*Astroscopus*), showing the areas behind the eyes where the electric organs (modified eye muscles) are housed; (d) the electric catfish (*Malapterurus electricus*), electric organ stippled; (e) the electric eel (*Electrophorus electricus*); (f) a mormyrid; (g) a knife-fish (*Eigenmannia*). Electric organs shown black in e, f, and g.

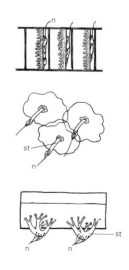

37 Innervation of electroplaques. Above: three electroplaques showing diffuse innervation and infolding of the opposite non-innervated face. Middle: three electroplaques of the electric catfish showing innervation through a stalk. Bottom: a mormyrid, innervation through branched stalks (*st*); nerve (*n*).

parts of the axial muscles. But the organs of torpedo-rays come from the hypobranchial muscles, and those of stargazers are modified parts of the eye muscles.

Of all the electric fishes, convergences between the mormyrids and gymnotids are the most intricate. The former, which are placed in a suborder of the Isospondyli, live in African fresh waters. The gymnotids, belonging to the cypriniform division of the order Ostariophysi, live in fresh waters of Central and South America. Both groups are weakly electrical, and it is now clear that they use their electric fields to locate obstacles and living organisms, particularly their prey and predators. They can also recognize the discharges of members of their own kind, and perhaps those of related species.

The electroreceptors are modified lateral-line organs, which are set in a specially elaborated epidermis. Above the basement membrane and innermost germinal layer, a layer of flattened cells is sandwiched between two layers of polygonal cells. Presumably this kind of epidermis provides adequate insulation between the electroreceptors. Many of these receptors are what Szabo (1965) calls tuberous organs, and each is housed in an evagination of the basement membrane. In gymnotids these sensory cells rest on a hillock of supporting cells and are covered by special epithelial cells. But the sensory cells of mormyrid organs are larger and each has its own cavity (Figure 38). Both groups also have ampullary organs[6] very similar to those of elasmobranchs and certain catfishes. These organs are also sensitive to small electric charges but their precise part in the lives of gymnotids and mormyrids has still to be discovered.

Central coordination of electroreceptive response seems to reside in a massively enlarged part of the cerebellum, the valvula. But accurate location of distortions in the electric field, whether caused by obstacles or organisms, depends on the maintenance of field symmetry, which is only possible if a fish keeps all or most of its body straight as it swims (Lissmann and Machin, 1958). Gymnotids keep such a posture by undulating their long anal fin, and so can certain mormyrids with a reduced caudal, which ripple their long-based dorsal and anal fins. Mormyrids with a well-developed caudal fin are stiffened in

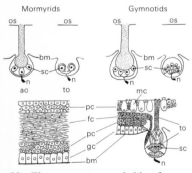

38 Electroreceptors and skin of electric fishes, mormyrids on the left, gymnotids on the right. Above: ampullary organs (*ao*) and tuberous organs (*to*); *os*, outer surface of skin; *sc*, sensory cell; *n*, nerve fibre. Below: sections through the skin of a mormyrid (*Marcusenius*) and a gymnotid (*Hypopomus*). Note the correspondence between the two in the layers of cells; *pc*, polygonal cells; *fc*, flattened cells; *gc*, germinal layer; *bm*, basement membrane. The section through the skin of the gymnotid also shows mucous cells (*mc*) and a tuberous organ. (Redrawn from Szabo, 1965.)

6. Mormyrids have a third special kind of lateral-line organ, which Szabo (1965) calls Type B. All three types have their counterpart in *Gymnarchus*.

110

the region of their electric organs by the four longitudinal (Gemmingers') bones that are inserted over and above these organs. In evolving toward these postural ends, members of the two groups have converged in body form and fin patterns (Lissmann, 1958). Indeed, Lissmann also draws attention to other convergent features, such as the development of long, elephantine snouts.

Electric fishes thus converge at levels of organization represented by organelles, cells, tissues, and organ systems. There are also behavioral convergences. But the evolution of electric organs is not always a response to common ecological conditions. Torpedo-rays and skates, which live and feed on the sea floor, have, respectively, powerful and weak electric organs. Torpedo-rays evidently use their electric organs to stun their prey and, no doubt, to deter their enemies. The development of the electric organs of skates varies considerably from species to species. Ishimaya (1955) believes that the organs of some skates guard them against attack from the rear. Thus, species with the smallest organs tend to have the tail wall armed with spines, whereas those with the largest organs have weakly spined tails. But if the convergences between the mormyrids and gymnotids are the most far-reaching, so is the nature of their surroundings. The members of both groups live in tropical waters that are often turbid. Moreover, Lissmann (1961) remarks that all the gymnotids are nocturnal, which is also true of many mormyrids. Certainly, the eyes of both are small; they have evolved means of electrolocation that have a greater range than the lateral-line system. One may even wonder whether some of the mormyrids and gymnotids have been able to evolve long, probing snouts because of their accurate location of prey. The olfactory organs of mormyrids at least are quite small. And other fishes with long snouts, such as pipe-fishes, seahorses, certain butterfly-fishes, and so on, have very well-developed eyes.

There is even convergence between the patterns of electric discharge. *Gymnarchus* and some gymnotids, such as *Sternarchus*, *Eigenmannia*, and *Sternopygus* produce continuous sinusoidal pulses of great regularity and high frequency. Mormyrids and certain gymnotids (*Hypopomus*, *Steatogenys*, and *Gymnotus*) produce brief repetitive discharges of relatively low frequency, which can be modified by external stimuli (Lissmann and

Schwassmann, 1965). These authors assume that the first type of discharge gives a fish better information of its environment. Earlier, Lissmann (1961) concluded that gymnotids producing this type tend to live in briskly flowing waters. Those producing brief, polyphasic patterns are sluggish, bottom-dwelling forms.

Convergence of organ systems

THE SENSORY SYSTEM

Eyes. Freshwater fishes range from brilliantly lit to pitch-dark surroundings. The same is true of marine fishes, except that the sunless reaches of the ocean, at least to a depth of 2,500 meters, are intermittently illuminated by sparks of bioluminescence. Moreover, as sunlight penetrates water, the parts of the spectrum are differentially absorbed, so that the twilight levels are lit by blue-green rays. It would not be surprising, then, to find some convergences in the eyes of fishes.

Because water has so high a refractive index that the cornea is optically "absent," the ancestor of fishes presumably had a spherical crystalline lens that bulged well through the pupil. Water is also relatively opaque to light, so this ancestor is likely to have possessed highly sensitive visual cells (rods) and perhaps cells (cones) for vision during the hours of daylight. But, whatever the precise ocular design of the ancestral fish, there have been many variations on this design.

In the outer covering and molding of the eyes there is considerable convergence. Secondary spectacles, those that shield the eye and develop from the cornea are present, for instance, in fishes that live near or on the bottom for example, catfishes, loaches, eels, gobies, cottids, cling fishes, and so forth. Walls (1942) suggests that spectacles protect the eyes from mud and sand stirred up from the bottom, as during feeding. He also observes that secondary spectacles are found in fishes that spend part of their life out of water (for example, mud skippers and climbing perch).

Diverse groups of teleosts have "adipose" eyelids. These eyelids, which are rarely adipose, cover the circumocular sulcus, so as to smooth off the bulge of the eyeball and they are transparent where they overlap the cornea (Figure 39). They are found, for instance, in *Hiodon* grey mullet, and certain clupeids, salmonids, and carang

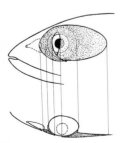

39 The head of a mackerel showing the "adipose" eyelids (stippled). (After Walls, 1942.)

gids. All such fishes have a shapely, streamlined body, capable of rapid movement. During bursts of speed the eddies stirred around a bulging eyeball presumably exert a considerable drag: hence the need for eyelids that smooth the run of water over the eyes. The eyeball must, of course, project so that the lens can collect light over a wide field. In the end then, the independent evolution of adipose eyelids seems to be related to the high refractive index of water. But in Pacific herring, Stewart (1962) found that the tissues of these eyelids is birefringent and that the amount of light absorbed varies with the plane of polarization. Whether these herring can detect, for instance, variations in the amount of polarized light during the day, remains to be shown. At all events, convergence in the design of adipose eyelids may be very close, as between those of *Pomolobus*, a clupeoid, and *Rastrelliger*, a scombroid (Walls, 1942).

We have seen that biochemical convergences between eyes involve their visual pigments. Concerning the retina itself, much could be said of the independent evolution of diurnal and nocturnal types (Walls, 1942). Reference is made above to the two types of mosaic formed by single and double cones in teleosts. Square mosaics are characteristic of percomorphs. But in some kinds, such as *Stizostedion* and *Lucioperca*, whose eyes are adapted for vision in rather dim light, rows of cones are present. On the other hand, some clupeids and salmonids have square mosaics over part of the retina. For instance, there are squares over the ventral part of the herrings' retina, which is adapted for acute vision, and rows over the dorsal part. Multiple cones evidently allow cone vision to persist in dimmer surroundings than are possible with single elements. Square mosaics seem to be particularly suited for the perception of movements (Engström, 1963).

Convergences in the eyes of deep-sea fishes will be evident from earlier discussion. Mesopelagic forms such as the stomiatoids and myctophids and slope-dwelling, benthic, or benthopelagic kinds (for example, macrourids), all have relatively large dioptic parts that collect and focus dim light on to a retina of long, closely packed rods. Instead of long rods some species have several layers of rod acromeres. This kind of retina has evolved independently in certain stomiatoids, argentinoids, and *Stylophorus*. The overall thickness of the acromere layer, whether single or multiple, is greatest over the central part of the fundus is laterally directed eyes. In eyes adapted for greater binocular vision, a special, temporal part of the retina in normally formed eyes, or the main retina in tubular eyes, has the longest acromere layer. Evidently, the longer the light-absorbing parts of the rods, the greater their sensitivity. Thus, to see effectively at twilight levels in the ocean, it is hardly surprising that such adaptations have been independently evolved several times.

Tubular eyes have evolved convergently in at least eight distinct groups of mesopelagic fishes. The structure and probable functions of these eyes, and their ecological significance are discussed in Chapter 3. But one remarkable kind of convergence was not considered. Ocular diverticula are present in the distantly related groups, Opisthoproctidae and Giganturidae (see Brauer, 1908; Munk, 1966). In *Bathylychnops*, a member of the first group the diverticulum develops below the anterior half of each eye, which looks upwards at an angle of about 35° to the perpendicular. The diverticulum is formed from the wall of the eyeball and contains all the essential parts of an eye, including a lens derived from the sclerotic layer. This small eye is provided with muscles, and judging from photographs, it can be moved over an arc below each main eye. In the other opisthoproctids and *Gigantura* the diverticulum is a smaller and simpler derivative of the retina. It contains a layer of short rods, backed in part by pigment so as to leave a window. Since these rods seem to be completely formed, the diverticulum could record luminescent events outside the highly specialized, binocular field of vision. The small diverticular eyes of *Bathylychnops* might even follow the movements of luminescent organisms. Yet other deep-sea fishes with tubular eyes manage well enough without retinal diverticula. Even so, the independent evolution of diverticula by opisthoproctids and *Gigantura* suggests some common functional value.

A further remarkable ocular convergence in mesopelagic fishes is the development of a temporal fovea in certain bathylagids, searsids, and scopelosaurids. In these fishes binocularity is increased by an aphakic space before the lens. Over the fovea, the light-absorbing parts of the rods are considerably longer, and presumably more sensitive, than those over the rest of the retina. The

more usual kind of cone fovea has also arisen independently in a number of shallow-water marine fishes,
notably in sea horses and pipe fishes, certain serranids,
rudder-fish (*Girella*), wrasses (*Julis*), and blennies. In two
trigger-fishes (*Balistes capriscus*, *Balistapus aculeatus*),
a puffer-fish (*Tetraodon fluviatilis*), and weever-fish
(*Trachinus*) the foveae are not so well formed. Walls
(1942) observes: "In all of the above species except the
sea horses, the fovea is located strongly temporally in the
retinal region that can see binocularly. But while these
fishes can and do converge their eyes to aim both foveae
at a prey object, the eyes are moved independently and
are not conjugated, but only co-ordinated temporarily in
each act of convergence. In fact such fishes are the only
vertebrates which can employ a temporal fovea for
monocular vision."

Any fish with an aphakic space before the lens is likely
to have a temporal fovea or a specially developed temporal part of the retina. Besides the foveal fishes listed in
the last paragraph (but not the sea horses and pipe fishes),
many others have eyes with an oval pupil and a forward
aphakic space, for example, pike, chlorophthalmids,
Aulostomus, certain angelfishes, cottids, ogcocephelids,
and so on. Though pike are known to have a temporal
fovea, study of retinal adaptations in the other forms has
yet to be made.

The acoustico-lateralis system. The ancestor of fishes
presumably had ears with semicircular canals and gravity
receptors. We might even suppose that the maintenance
of muscular tone was also a primitive function of the ears,
but could they also receive sounds? At all events, the
semicircular canals of the earliest fishes were probably
just as properly adapted to their form and motion as are
those of modern species. It is well known that the semicircular canals respond to angular accelerations, but
appreciation of how nicely these canals fit the design of
fishes is due to the work of Jones and Spells (1963). Compared to mammals, birds, and reptiles, fishes have a
wider internal radius and a greater radius of curvature of
the semicircular canal. Jones and Spells imply that fishes
need such canals because they make relatively slow turns.
Fishes have no neck, of course, and in turning are opposed by a dense medium. Hence, there is a need for extra
sensitivity to angular accelerations, which can be gained
by having wide semicircular canals. For instance, the

average internal radius of the vertical canals of fishes (0.23 mm) is nearly twice the corresponding figure for mammals.[7] Such increase in canal sensitivity calls for a correlated increase in the time constant of cupular return, which Jones and Spells say can be got by a more than proportionate increase in the radius of curvature.

Some fishes, of course, turn more slowly than others. Eel-shaped fishes, as Gray (1933) showed, change direction more slowly than fusiform species. Using Jones and Spell's arguments, we might expect eels to have extrawide semicircular canals. Certainly their figures show this is true of the conger eel. Reference to accurate figures of the ears of other eels also confirms this prediction. More widely, the ears of eel-shaped fishes should converge in having wide-bore canals with a high radius of curvature. This again, is true, judging from good figures. For instance, *Idiacanthus*, an eel-shaped stomiatoid, has these extra dimensions when compared to its fusiform relatives. Presumably the wide vertical canals of lampreys are also correlated with their form.[8]

"Flatfishes" of comparable size also have extrasensitive semicircular canals, as may be seen by looking at Jones and Spell's figures for the Greenland halibut and the common skate (inner radius of the vertical canals is 0.30 mm and 0.35 mm, respectively). Again, these extra dimensions seem "right" when we think of the shape and motion of these two fishes and the slowness of turns that are monitored by their vertical canals. Fishes with a long tapering tail do not have the means to make quick turns. It seems apt, then, for the rabbit-fish (*Chimaera monstrosa*) to have wide vertical canals (inner radius 0.315 mm), with a large radius of curvature. The same is probably true of the macrourids. To summarize, fishes with certain patterns of form and motion have converged in those dimensions of their semicircular canals that control their sensitivity to angular accelerations.[9]

7. Jones and Spells' figures show clearly that in fishes the internal radius of the horizontal semicircular canal is smaller than that of the vertical canals. This feature of the fish ear does not seem to have been noticed. Presumably there is less need for sensitivity to the relatively quicker turns made about the vertical primary axis of fishes.

8. Are the wide semicircular canals of sea horses related to their special mode of propulsion, which only permits very slow turns?

9. One should, of course, compare species of much the same size, for the larger the fish, the greater its need for extra sensitivity. Thus the inner radius of the vertical canals of a 9.05 kg Greenland halibut is almost twice the same dimension in a 0.34 kg lemon sole.

Aspects of Convergent Evolution

The ancestral fish presumably had some kind of lateral-line system. In modern fishes the sensory units (neuromasts) are either housed in mucous-charged canals or freely exposed to the water. Neuromasts, which consist of hair-bearing sensory cells capped by a gelatinous cupula, are displacement receptors. They respond to nearby disturbances, such as are caused by the movements of prey, predators, and social partners. The neuromasts are so disposed and integrated that a fish is able to find a moving form, and even to gain some impression of its size, velocity, and direction.

If the lateral-line organs "told" fishes about the large-scale movements of their environment, there might well be considerable convergence in their design. But neuromasts have no true rheotactic function. There is, however, some correlation between the type of neuromast and the motion of water. Thus, very active fishes, or those that face lively currents, have the neuromasts in canals. Free neuromasts would be too "noisy." Canals tend to disappear in fishes that swim slowly, or intermittently, or live in quiet surroundings. Thus in slowly moving bathypelagic waters, ceratioid angler-fishes, gulper eels, and a true eel (*Cyema*) converge in developing free neuromasts only, which are carried on stalks. In his review, Dijkgraaf (1962) observes that canals are absent in syngnathids, a loach (*Misgurnus fossilis*), sticklebacks, various gobies, the angler-fish (*Lophius*), and the South American lungfish. He concludes: "All are typical slow swimmers or bottom dwellers, mostly living in quiet waters like the amphibians, which are also devoid of canals."

If there is a correlation between intermittent movement and type of neuromast, larval teleosts show it well. Whether they are going to develop canal organs or free-ending organs or both, the larval neuromasts project slightly from the skin and bear rod-like cupulae. But the functions of these larval neuromasts are not yet quite clear. Conceivably, they could enable larvae to evade enemies and catch food. At all events, larval cyprinids that live in quiet, weedy surroundings have more free neuromasts than the salmonids that spend their early life in gravel washed by brisk streams.

Dijkgraaf (1962) has noted two main orientations of the neuromasts in canals: in diverse fishes with narrow canals the long axis of the sensory hair tract (and of the

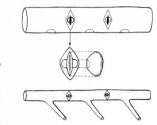

40 Two main types of lateral-line canal organs. Above: wide canals with large neuromasts having the functional axis at right angles to the canal axis. Below: narrow canals with tubular exits, having small neuromasts with the functional axis along the canal axis.

The Life of Fishes

cupula) is parallel to the axis of the canal. This is also true of the burbot (*Lota*), which has rather wide canals, but at the level of each neuromast the canal is narrowed by connective tissue (Flock, 1965). Fishes with very wide canals, such as macrourids, *Notopterus,* the ruffe (*Acerina*), and the pirate perch (*Aphredoderus*), have the neuromast axis at right angles to the canal (Figure 40). Doubtless, these two sets of convergences have a functional significance. Fishes with wide canals have very large neuromasts. Indeed, in a series of macrourids I find that there is a direct relation between the area of a particular neuromast and the volume of the relevant section of the canal. Since neuromasts respond only to disturbances along their axis, large and very sensitive elements are presumably well buffered against disturbances due to water passing along the moving fish (Marshall, 1965a). Wide canals are often closed; narrow canals are generally open to the water, so that current-like disturbances at right angles to the surface result in a flow of mucous along the functional axis of the neuromasts. Moreover, many fishes have evolved labyrinthine outlets of their main canals. But the functional significance of such development in clupeids, horse-mackerel, flying-fishes, and so forth, is not understood (Dijkgraaf, 1962).

Chemoreceptive systems. Apart from their common chemical sense, which resides in nerve endings in the skin, fishes receive chemical intelligence by way of their olfactory organs and taste buds. The olfactory organs enable a fish to find a distant source of smell, but tasting depends on contact. Recently, though, Bardach, *et al.* (1967) have shown that the catfish *Ictalurus nebulosus* can find distant food through its gustatory system alone.

The essence of olfactory design is some means whereby water is circulated over the sensory epithelium. There are three main ways in teleosts, each of which has been independently acquired by diverse groups. In pike, pipe fishes and certain puffer fishes, for instance, the nostrils are so disposed that water is deflected through the olfactory capsule as the fish swims. (But these are all microsomatic fishes.) Diverse fishes that rest on the bottom or spend much of their time in a hovering position, have evolved an aspirator kind of mechanism. In such kinds as the bullhead, *Myoxocephalus scorpius*, the viviparous eel pout (*Zoarces viviparus*), sticklebacks, and certain

118

flatfishes (for example, *Scophthalmus maximus*), breathing movements relax and squeeze an olfactory sac to circulate water in and out of a single nostril. The third main kind of circulatory mechanism depends on breathing movements or ciliary currents or both, whereby water is drawn through an anterior nostril, over the olfactory epithelium, and out of the posterior nostril.

Convergence between the main parts of the olfactory system is well seen in the table given by Burne (1909, pp. 656–657). For instance, a rosette with longitudinal lamellae is found in *Ophicephalus* and flatfishes (*Hippoglossus* and *Pleuronectes*)—and also in angler fishes (Pediculati). Each nasal capsule has a posterior nasal sac in various isospondylous fishes (*Salmo, Osmerus, Coregonus, Clupea*), sticklebacks, barracuda, hake, and mackerel.

Concerning the gustatory system, we have already referred to Freihofer's (1963) work on the patterns of the ramus lateralis accessorius nerves. This complex innervates taste buds on the external surface of the head and body, excluding the snout. The detailed pattern of RLA nerves is certainly a valuable indicator of relationships. For instance, pattern 9 is typical of percomorph fishes, and pattern 7 is found in cod-like fishes, hake, and macrourids (Anacanthini). But the development of RLA pattern 15 by ophidioid fishes and eel pouts (Zoarcidae) may well be convergent and related to similar body-form, fin patterns and habits. This may also be true of the cottids and cling-fishes, which are classed in different orders but have RLA pattern 12. In any event, there can be no doubt that the extension of the gustatory and tactile senses through the development of barbels is a classic instance of multiple convergence. One has only to think of cyprinids, catfishes, codfishes, macrourids, polynemids, red mullet, armored sea-robins, and so forth.

CENTRAL NERVOUS SYSTEM

Although there are convergences at the cellular level in the central nervous system of fishes[10] we are here con-

10. Notably in neurosecretory systems. For instance: "Both structural and cytological characteristics of the caudal neurosecretory system indicate that its evolutionary history among teleosts was diverse; similar advanced features appear to have evolved independently in separate phyletic lines" (Fridberg and Bern, 1968).

cerned only with grosser features. Convergences in the olfactory and cerebellar systems spring most readily to mind. In vertebrates, fibres from the olfactory cells make glomerular synapses in the olfactory bulbs, and from the bulbs nerve fibres pass to olfactory centers in the forebrain. In most fishes the olfactory bulbs develop just ahead of the forebrain, but in ostariophysan and most anacanthine teleosts the bulbs are close to the olfactory epithelium. This is clearly convergence, and one may wonder whether the similarity is related to some special olfactory function, common to these two orders of fishes. We have no such physiological knowledge, but the "explanation" may be thus. Olfactory functions are not dependent on the position of the olfactory bulbs; these are simply correlated in development with either the forebrain or the olfactory epithelium. In the adult, though, the bulbs may lie between the forebrain and the olfactory epithelium (for example, in *Merluccius* and *Bathygadus* among the Anacanthini).

Convergences in cerebellar structures are most evident in the valvula, the part that lies under the optic tectum. In mormyrids and gymnotids the valvula is enormously expanded, largely it seems to serve as a center for the many electroreceptors (modified neuromasts) in the skin. More generally, the valvula is thought to control orientation through the acoustico-lateralis and gustatory systems. Tuge, Uchihashi, and Shimamura (1968) state that teleosts with a well-developed corpus cerebelli also have a large valvula. Here, though, we will only consider the elaboration of the valvula and in particular the patterns of the nuclear and molecular layers as seen in sections. A simple, sac-like form of valvula is found in gadids (Figure 41), synodontids, gasterosteids, syngnathids, poeciliids, uranoscopids, blenniids, ophidiids, callionymids, gobiids, cottids, pleuronectids, soleids, and gobiesocids (Bănărescu, 1957). Now, gadids have a large cerebellum, but in most of the other groups this organ is relatively small, and the simplicity of their valvula is just part of their limited cerebellar development. Hence the "convergences" between the above groups. In teleosts with a more complex valvula cerebelli, those that Bănărescu groups under the carangid type are as follows: Esocidae, Apogonidae, Theraponidae, Serranidae, Carangidae, Maenidae, Sparidae, Chaetodontidae, Pempheridae, Pomacentridae, Labridae, Zanclidae, Anaban-

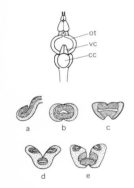

41 Patterns in the molecular and nuclear layers in the valvula cerebelli of certain teleosts. (Nuclear layer has heavier dotting.) Above: a diagram of the brain with the optic tectum (*ot*) removed showing the valvula cerebelli (*vc*), corpus cerebelli (*cc*). (a) and (b), longitudinal and transverse sections through the valvula of the burbot, *Lota*; (c) to (e), transverse sections through the valvula of (c) *Blennius sanguinolentus*, (d) *Esox lucius*, (e) *Apogon imberbis*. (Redrawn from Bănărescu, 1957.)

tidae, Scombridae and Triglidae. Morphologically, the carangid type of valvula can be derived from the salmonid type, which has a "dorsal sac" and a "ventral sac," through the division of the dorsal sac into two lateral lobes. Here the patterns of the nuclear and molecular layers (Figure 41) are good indicators of relationships, but it is odd to find the pike (*Esox*) among so many kinds of percomorph fishes. The convergence is presumably related to this constraint: there are few structural ways of elaborating the valvula. The pike has simply "chosen" a pattern common to many spiny-finned fishes. Similarly, the belonids have "copied" those percomorphs with a centrarchid type of valvula.

THE TELEOST SWIMBLADDER

During the embryonic life of bony fishes the swimbladder develops as an outgrowth of the foregut. In lungfishes, bichirs, holosteans (*Amia* and *Lepisosteus*)—and certain primitive teleosts—this outgrowth becomes an air-breathing organ. But in nearly all teleosts it is essentially a hydrostatic device (Jones and Marshall, 1953). The ancestral teleost presumably had a swimbladder opening through a pneumatic duct into the foregut, and during the early radiations of the group, this open organ was inherited by the lineages now represented by isospondylous, ostariophysan, and apodal fishes. Convergences in the swimbladder of these forms largely involve the design of gas-secreting tissues.

Fishes with an open swimbladder (Figure 42) inflate or deflate it through the pneumatic duct. When such a fish moves toward the surface it has an efficient means of losing gas, but the acquisition of enough gas by gulping air at the surface poses physical problems. For if a swimbladder is to function properly as a hydrostatic organ, its buoyant volume must be kept fairly constant at ambient water pressures. Even if we suppose that a fish is able to gulp enough air to give it neutral buoyancy at the depth of its living space, the fish must overcome the uplift due to its positive buoyancy as it swims down. After this, any appreciable increase in its depth would require it to revisit the surface. Clearly the deeper the living space the greater the physical stress. Is it so surprising then, that most marine teleosts have evolved a closed swimbladder

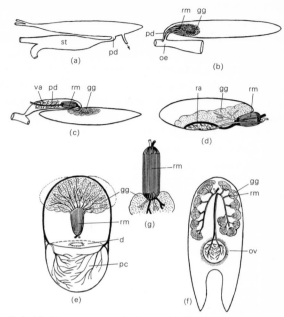

42   Swimbladder structure of teleosts. (a) Herring (*Clupea*) with the
pneumatic duct (*pd*) opening into the pyloric end of the stomach (*st*).
The canal behind the pneumatic duct opens near the anus. (b) Pike
(*Esox*), pneumatic duct opening into the oesophagus (*oe*). At the
anterior end there are retia mirabilia (*rm*) and gas gland cells (*gg*).
(c) An eel (*Anguilla*) showing the vascularized, resorbent area (*va*) of
the pneumatic duct, the two retia mirabilia in the walls of the duct
and the gas gland in the main chamber. (d) A stomiatoid, showing
resorbent area (*ra*) and the single, posterior rete mirabile. (e) *Anten-
narius multiocellatus,* a frogfish. The forward chamber of the swimbladder, containing the rete mirabile and the gas gland, is separated
by a diaphragm (*d*) from the vascular, resorbent posterior chamber.
(f) *Acanthurus chirurgus,* a surgeonfish. Resorbent area, an oval (*ov*)
which receives its blood vessels from the same system that supplies
the horse-shoe shaped tract of glandular tissue. (g) Diagram of a
bipolar rete mirabile showing the parallel system of venous
capillaries (black) and arterial capillaries (white).

equipped with gas-secreting structures? Moreover, these
structures are also found in diverse species that have an
open organ, and most of these, being eels, are also
marine.

In evolving retia mirabilia and special glands for
secreting gases into the swimbladder, teleosts are unique
among modern fishes.[11] A *rete mirabile*, better called a
fascis mirabilis, is a close association in parallel of arterial
and venous capillaries (Figure 42), and those in the swim-

11. Certain teleosts lacking retia are still able to secrete gases
(Fahlen, 1967).

bladder circulate blood through the gas glands. The entire retial system has two functions, both depending on its many micro-countercurrents of arterial and venous blood. Exchange of gases between the outflowing and inflowing blood maintains tensions within the swimbladder. Countercurrent multiplication of gases during successive circulations of blood builds tensions high enough to overcome the pressure gradient between gases in the inflowing blood and those in the swimbladder. The gas gland, which develops from the inner epithelium of the swimbladder, is thus presented with gases at appropriate tensions.

These elaborate gas-producing systems have evolved independently within several groups of teleosts. In those with an open swimbladder they occur in certain salmonids (*Coregonus*), pike (*Esox*; Figure 42), cyprinids, elopoids (*Albula*), and the eels. Of these, *Albula* and the eels probably inherited their retia from a common elopoid ancestor. Both have two bipolar retia associated with the pneumatic duct, which in eels becomes highly vascularized and acts as a resorbent organ (Figure 42). Moreover, certain lines of physostomatous teleosts evolved deep-sea representatives that converge in forming both a closed swimbladder and gas-producing systems. In the albulid, *Pterothrissus*, and the synaphobranchid eels, which belong to the elopoid line, the pneumatic duct is present but closed. The stomiatoids, evidently most closely related to the osmeroids (Weitzman, 1967), have lost all trace of the pneumatic duct but converge with the eels in developing bipolar retia mirabilia (Marshall, 1960). The argentinoids are closest to the salmonids, and like the *Coregonus*, have micro-retia mirabilia. But they have lost the pneumatic duct; instead they evolved a highly vascular, resorbent posterior chamber, which in *Argentina silus*, at least, is connected to the anterior, gas-producing chamber through a muscular sphincter, able it seems to close the connection (Fahlen, 1965). In developing this resorbent system, the argentinids converge with several groups of "higher" teleosts, for instance, with some of the percomorphs.

In the rest of the teleosts the swimbladder is closed (physoclistous) and there is no trace of a pneumatic duct in the adult. Again, a closed sac and, presumably, a gas-secreting complex, have been acquired independently several times, for instance, by the Myctophidae and *Neoscopelus* among the Iniomi; the Miripinnati, gadoids, percopsids, atherinomorphs, pediculates, and spiny-finned groups (Percomorphi and Scleroparei). Here we need consider only the resorbent system, which is either euphysoclist or paraphysoclist. The first type of swimbladder is deflated through an oval or posterior chamber (Figure 42), but in paraphysoclists the resorbent capillary network is not so sharply separated from the gas-producing tissues. This second system is found in flying-fishes, and so on (Synentognathi), killifishes, and so on (Microcyprini), and silversides (Atherinidae), and this may well provide more support for placing these fishes in a single group, which Rosen (1964) calls the order Atherinomorpha. For, if we argue that this similarity between these groups is convergently related to their surface-seeking life, and thus perhaps to modest need of a resorbent system, we must not forget that dominant deep-sea fishes, the stomiatoids, are paraphysoclists. Moreover, some stomiatoids visit the surface layers each night (Marshall, 1960), which implies that the paraphysoclist resorbent system is not necessarily less efficient than the euphysoclist.

Concerning the resorbent system of euphysoclists, both oval and posterior chamber have these common features: (1) the capillary network is set very close to the swimbladder gases; (2) circular and radial muscles control the opening between the gas-producing and resorbent parts; and (3) between these two parts there is a sharp change in the structure of the swimbladder wall. Where a diaphragm rather than a muscular sphincter separates anterior and posterior chambers, the former not only has circular and radial muscles, but other muscles may move it forward or backwards (Fänge, 1953). But oval and posterior chambers are essentially homologous. Indeed, Fänge (1953) found either an oval or a diaphragm in the swimbladder of the burbot (*Lota lota*).

It is usual among euphysoclists for major groups to "choose" either an oval or a posterior chamber. For instance, an oval has evolved independently in Myctophidae, Anacanthini (except *Gaidropsarus*), Melamphaidae, Percidae, Gobiidae, and Balistidae. A posterior chamber is developed, again independently and *inter alia*, by percopsids, antennariids, ophidioids, holocentrids, zeomorphs, Scleroparei, and serranids. Functionally, there seems to be no advantageous difference between an oval and a posterior chamber. Of the mesopelagic fishes, both the myctophids and certain of the gempylids (for example, *Nealotus*) swim upward each day to the surface layer. The swimbladder of the myctophids has an oval, and that of the gempylids has a posterior, chamber. And the myctophids, at least, judged by observations from underwater vehicles, are neutrally buoyant, or nearly so, at their daytime levels. Those with a swimbladder thus start their upward migration with a volume of gas equal to about 5 percent of their body volume and at a pressure corresponding to that of their surroundings.

ON CERTAIN FEEDING DEVICES

During the evolution of fishes, jaw mechanisms became more versatile. Teleosts, which acquired the most mobile jaws, have been the dominant fishes since late Cretaceous times. Moreover, many of them evolved protrusible jaws, certain advantages of which are considered elsewhere. Such advantages are by no means fully appreciated, but they are evidently considerable. Cyprinids and percomorphs, both with protrusible jaws, are outstandingly successful fishes. Some of the atherinomorph fishes (atherinids and cyprinodonts) also evolved such jaws. Now these three groups are remotely related, but their jaw mechanisms are alike in these respects (Alexander, 1967a): (1) the premaxillae exclude the maxillae from the gape; (2) the pedicels of the premaxillae are attached to and overlie a median mobile cartilage or ossified element (kinethmoid); (3) the maxillae are also attached (by ligaments) to this mobile element; (4) the head of each maxilla bears a forward, medially directed process and a condyle moving on the ethmoid region of the neurocranium; and (5) a superficial part of the adductor mandibulae muscle inserts on the maxilla. Though we are uncertain how much of this resemblance was derived from a remote common ancestor, it is most probable that the protrusible mechanisms as a whole were independently evolved by each group. Moreover, the precise mechanism of protrusion is different in all three. In the percomorphs, as the maxilla turns inward a condyle on its head acts like a cam on the articular pro-

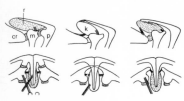

43 Three kinds of protractile jaw. Above: side views; below: dorsal views. Left: spiny-finned (percomorph); middle: cyprinid; right: atheriniform. Cartilage is stippled. *k*, kinethmoid: *r*, rostral cartilage; *p*, premaxilla; *m*, maxilla; *cr*, cranium. (Redrawn from Alexander, 1967.)

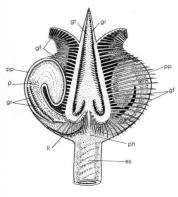

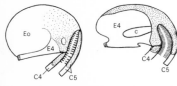

44 Epibranchial organs. Above: of *Chanos* the milk-fish; ventral view showing the right sac cut open to reveal the continuation of the gill rakers (*gr*) of the fourth gill arch into the organ. *gf*, gill filaments; *pp*, pharyngeal pocket; *p*, papillae; *ph*, pharynx; *es*, esophagus; *ll*, lower limb of fourth gill arch. (Redrawn after Kapoor, 1954.) Bottom: plan of epibranchial organs of an osteoglossid, *Heterotis niloticus* (left), and a clupeid, *Brevoortia tyrannus* (right). Cartilage is stippled. *Eo*, epibranchial organ; *E4*, fourth epibranchial; *C4* and *C5*, tips of fourth and fifth ceratobranchials. (Redrawn from Nelson, 1967b.)

cess of the premaxilla. The cyprinid mechanism involves the rotation of the median kinethmoid bone about its transverse axis, though one end of this bone is attached to the cranium (Figure 43); Alexander, 1967). Study of convergent features reveals that they rarely involve the closer similarities that are due to common inheritance.

Even so, most species of another successful group of fishes, the macrourids, have protrusible jaws essentially like those of percomorphs. Presumably the common ancestor of the macrourids and their relatives (codfishes, hake, and so on), which have fixed jaws, possessed jaws very like the ancestor of percomorph fishes. Given such jaws, there was only one possible way, as it were, for the macrourids and percomorphs to evolve protrusible jaws. Still, the end products are extraordinarily alike.

Convergent features may even mimic those that owe their similarity to common inheritance. Is this true of epibranchial organs, which occur in certain microphagous fishes? These organs are large and elaborate in *Heterotis niloticus* (Osteoglossidae), certain characids, clupeids, engraulids, and in *Phractolaemus* and *Chanos* (Figure 44). Small organs develop in kneriids, *Gonorhynchus*, *Alepocephalus*, and some of the clupeids and engraulids. The common features of epibranchial organs are these: (1) the association of the epibranchial elements of the fourth and fifth gill arches; (2) the upward extension behind the fifth epibranchial of the last gill slit and its rakers, which generally proliferate or are reduced (in characids); (3) the development of a cartilaginous support for the posterior row of gill rakers; and (4) the enlargement (except in *Heterotis*) of the fourth epibranchial, which supports from below a gradually expanding diverticulum of the pharynx (Nelson, 1967b; Figure 44).

Except in certain characids, prominent gill rakers in each epibranchial organ divide its lumen into two parts, one confluent with the gill chamber, the other with the pharynx. It looks then, as though water will stream through the epibranchial organs as a fish is feeding; thus small organisms will be screened by the rakers on the pharyngeal side of the organs. This food will be bound together by the mucous secretions of glands in the epithelium of the epibranchial organ, the walls of which are also very muscular (Figure 44). These muscles presumably squeeze boluses of mucous-trapped food into the esophagus. There is certainly some evidence (Miller,

1964) that phytoplankton and mud particles are retained by epibranchial organs. Moreover, as Nelson (1967b) says, fishes with highly developed epibranchial organs (for example, *Heterotis, Curimata, Chanos, Cetengraulis, Dorosoma*), which also have a "gizzard," feed on the finest kinds of food, such as phytoplankton and mud particles.

The alimentary structure of these fishes could well be convergent. A muscular, triturating gizzard is found also in gray mullet (Mugilidae), some of which feed on detritus, and in many species of the algae-browsing surgeonfishes. Have epibranchial organs, as Nelson believes, also been acquired independently? At first it seems curious that epibranchial organs have not been developed in some, at least, of the higher teleosts with microphagous habits (for example, certain cichlids and mugilids). But the gill arches of such forms are too specialized, particularly as mobile tooth-bearing bones, to form epibranchial organs (Nelson, 1967b). Perhaps the gill arches of their ancestors were also too specialized. Nelson advances the argument "that the potentiality for epibranchial-organ development was a similarity shared by, and from a systematic viewpoint tending to unite primitive members of the Osteoglossiformes, Cypriniformes, Clupeiformes and Gonorhynchiformes." This potentiality must have been inherent genetically, which would mean that epibranchial organs in these groups were the result of parallel evolution, canalized by the genetic constitution of their common ancestor. It is even possible that the ancestral teleost was microphagous and had epibranchial organs. Moreover, the structure of well-developed organs is so close that one may wonder whether the forms that have such elaborate food screening and processing devices acquired them convergently. There is also the problem of species with small epibranchial organs. Were these originally large? Indeed, a full survey of the lower teleosts may well reveal other species with large, small, and even reduced organs. In the end, we have to decide whether the teleosts are a monophyletic or polyphyletic group. Studies on the elaboration of the head lateralis system (Marshall, unpublished) could suggest a single ancestor. After all, the teleosts are so successful that one would be surprised if the genetic groundwork for such success could have evolved more than once.

ASPECTS OF CAMOUFLAGE

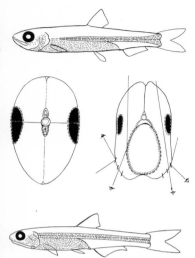

45   Silver-sided fishes. Above: a round herring, *Jenkinsia stolifera*. Below: an anchovy, *Anchoa lamprotaenia*. Silvery areas are stippled. Middle left: transverse section through the tail of *Jenkinsia* to show the silvery pigment (broken line) around the red muscle (solid black) and the silvering in the hemal canal of the vertebral column. Middle right: transverse section through the abdominal region of a silverside, *Atherina presbyter*. Silvery pigment (broken line) and black pigment (dotted line) indicated around the red muscle (black) and the body cavity. The bars on the lower flanks are orientated plates postulated to account for the reflection effects observed. (Redrawn from Denton and Nicol, 1966.)

Diverse forms of camouflage have risen independently in diverse groups of fishes. One has only to think of masking stripes that pass over the eyes and various types of color patterns, broken, mottled, striped, and so forth. Many groups converge in their means of changing color to match the shade of a substrate. Certain rays and diverse teleosts have developed one or more eye-spots (ocelli), which evidently mislead their enemies. Many fishes have independently come to resemble certain objects in their surroundings, particularly the parts of plants. The present intention, though, is to concentrate on one aspect of camouflage, shown by silver-sided fishes (Figure 45).

Atherinid fishes are called silversides. Most species have a silvery stripe along the flanks, centered on the division between the upper and lower parts of their myotomes. Silvery fishes usually have fully mirrored flanks. Their reflectivity has been studied closely by Denton and Nicol (1966), who have shown how mirrored sides tend to make a fish vanish into most fields of view. They also studied three silver-banded fishes: a smelt, *Osmerus eperlanus*, a silverside, *Atherina presbyter*, sometimes called the sand smelt, and a half-beak, *Hemirhamphus far*. Here is their appreciation of the second species: "The sand smelt is brown above, silvery on the sides and below, and has a prominent silvery band along the middle of the flank. Above the silvery band the translucent muscle shows through the skin.

"When a smelt is held so that light is falling on its dorsal surface from above, and the observer is viewing it obliquely from below, the lower sides and belly appear shiny and iridescent, blue or silvery or rose-coloured, depending on the angle of viewing. If the fish is now viewed from below, the body appears somewhat translucent, light passing through the body of the fish and emerging below. The silvery band on the flank then appears dark. With light striking the lower surface of the fish, and the observer viewing it obliquely from above, the lower flank of the fish appears silvery.

"There is very little silver on the scales of the flank below the lateral line and the silvery appearance of this region is not appreciably diminished when the scales are removed. Therefore, the silvering lies deeper in the skin and histological examination described below, reveals that there is a layer of oriented reflecting plates in the subdermis of the lower flank.

"The thin band of red muscle in each flank is invested by a thin silvery argenteum except for the dorsal surface which is black. The peritoneum up to the spinal region is covered with silver, and the inner surface is black. The sketch [see Figure 45] shows the postulated orientation of reflecting plates and layers to give the reflexion effects observed."

Certain other kinds of smelts (for example, *Hypomesus*), nearly all silversides and diverse species of half-beaks (for example, *Chriodorus, Hyporhamphus*) have this kind of color pattern. It is also displayed by certain clupeids (and round herrings), most of the anchovies (Engraulidae), and some needlefish (for example, *Strongylura acus*). Moreover, adult fishes with fully silvered sides may be silver-banded when young (for example, menhaden, *Brevoortia*, and thread herring, *Opisthonema oglinum*). To make a fuller survey, one would also have to include some of the characids (*Hyphessobrycon* spp.) and the glass-fishes (*Ambassis*, and so on).

All these fishes are partly transparent or translucent. But the opaque tissues "such as eyes, gut and red muscle are camouflaged by treating each as a unit which is dark above and silvery on the sides and below" (Denton and Nicol, 1966, p. 685). Red muscle fibres, in contrast to white, are opaque largely because of their more elaborate blood system. Here I should add that the skin over the pectoral muscles of certain fishes is silvery. Sweepers (*Pempheris* spp.), silver hatchet-fishes (Gasteropelecidae) and some of the silver perches (*Leiognathidae*), which have such silvering, make great use of their pectorals as paddles. Sustained movements of their pectorals evidently require the provision of red muscles.

Gills are also opaque, and in many silver-banded fishes the skin over the gill covers is silvery. But for this pigmentation, the gills would be readily seen through the thin opercular bones. These are also prominent blood vessels, the dorsal aorta and the caudal vein, in the hemal canal of the caudal vertebrae. Again, these blood vessels would be conspicuous but for the silver lining of the hemal canal. I have found such a lining in a round herring (*Jenkinsia stolifera*), an anchovy (*Anchoa mitchilli*) a half-beak (*Chriodorus atherinoides*), and a silverside

(*Menidia menidia*). (See Figure 45.) It must also be developed in other representatives of these groups.

The silvery disguise of opaque tissues has thus been evolved independently in six groups of teleosts at the very least. Considering the sets of opaque homologues to be so camouflaged, it is hardly surprising that the convergences look very much alike. Moreover, given the fact that these fishes spend some part of their life in well-lit surroundings and that they have "chosen" not to cover up their white muscle, their argentine patterns seem inevitable.

WAYS OF REPRODUCTION

In her recent book, Susanne Langer (1967) wrote thus: "In competition with other organisms of other kinds as well as its own, the 'more or less' of adaptedness is what matters: the less adapted will be reduced and gradually eliminated, leaving the better adapted to continue their lines. And, as heredity and mutation constantly produce new relative fits and misfits, there is a constant reviewing and culling of forms, which results in some really amazing adaptations; fishes that brood their offspring in their mouths . . . moths that sit on the bark of trees in such positions that their wing patterns line up exactly with the bark patterns and make them all but invisible, and countless other improbable but actual wiles." In fact, the "amazing adaptation" of oral incubation has evolved independently in osteoglossids (*Osteoglossum and Scleropages*), ariid catfishes (marine species), cichlids, apogonids, opisthognathids, and certain anabantids (*Betta picta* and *B. brederi*). Lastly, females of the cavefish *Amblyopsis* carry their eggs not in the mouth but in their gill chambers.

Oral incubation is thus practiced by both marine and freshwater teleosts. Male or female may be the incubator. For instance, male ariids carry the eggs, but in most species of the cichlid genus *Haplochromis*, it is the females. The male is the carrier in most apogonids, but both sexes of *Apogon semilineatus* have this ability (Breder and Rosen, 1966). The eggs are retained until they hatch, and the young usually remain with their parent until they are able to swim easily.

Diverse groups of fishes have converged in evolving the habits and means of producing a mass of heavy, adhesive eggs, which are generally placed in a nest or concealment and guarded by the male (for example, *Amia*, *Protopterus* spp., various catfishes, sticklebacks, sunfishes (Centrarchidae), gobies, cottids, blennies, and toadfishes). Apart from keeping predators away from the eggs, the male also aerates and cleans them by fanning movements of his pectorals. But these activities hardly indicate how the habit of oral incubation may have arisen. It will be better, though, to look at the closest relatives of mouth-breeders.

All the cardinal fishes (Apogonidae) are oral incubators, but some cichlids lay adhesive eggs, which may be placed in a nest. The parents sometimes probe the eggs or even mouth them and move them to another place. The jaws and mouth are, of course, the only available parts to do such work, which gives some inkling of how the habit of oral incubation may have arisen, particularly in response to heavy predation pressure on nested eggs.

Study of the labyrinth fishes (Anabantidae) is also helpful. Certain kinds produce floating eggs, but most of them place the eggs in bubble nests, which are usually blown by the males.[12] After the eggs are laid and fertilized—and they may be heavier than water—the male gathers them in his mouth and blows them into the bubble nest. This is the procedure for *Betta splendens*, but *B. brederi* and *B. picta* are mouth breeders. Breder (1934) observed that in *B. brederi* both sexes picked up the eggs, carried them to the surface, and then spat them upwards as though a bubble nest were there. After the eggs fell to the bottom, they were picked up again and the spitting behavior repeated. This sequence might continue for half an hour, after which, the male, assisted by the female, took all the eggs in his mouth. Again, it is easy to imagine how mouth-breeding habits took over from those associated with the building of bubble nests.

The valuable survey of reproductive modes by Breder and Rosen (1966) reveals other implicit convergences. Oral incubation is but one form of parental care, and each other form, such as viviparity, has evolved independently a number of times. Certain types of eggs, for instance, those with entangling or adhesive threads

12. Armored catfishes (*Callichthys callichthys* and *Hoplosternum littorale*) and a characid (*Sarcodaces odoe*) converge with the anabantids in making bubble nests.

are known in synentognaths, cyprinodontoids, atherinids, phallostethids, plesiopids, apogonids, blennioids, and gobiesocids. The pelagic kind of egg, largely buoyed by dilute body fluids, is widespread among groups of marine teleosts. The ancestral teleost may well have produced this kind of egg, but we are less certain that simple inheritance entirely explains its use by diverse groups. At all events, it is clear that groups depending predominantly on pelagic eggs are not precluded from evolving other kinds. For instance, the herring (*Clupea harengus*), two codfishes (*Eleginus navaga* and *Microgadus tomcod*), and certain flatfishes have independently acquired the habits and biochemical means for producing heavily yolked demersal eggs.

Some adaptive types of fishes

The adaptive radiations of organisms produce convergence as well as divergence. During the early evolution of a group there may well be much phyletic parallelism, but as divergence proceeds and adaptive zones are exploited, convergences will arise. But parallelism is always possible when genetic divergence between lineages of common ancestry has not gone so far as to exclude the canalized development of common characters in two or more of these lineages. It is, of course, only when such genetic limits have been passed that true convergences may arise.

Adaptive zones in the hydrosphere are centered on or near interfaces or at mid-water levels. Besides the air/water and land/water interfaces there are those between vegetation, rocks, corals, and so forth, and the medium. The earliest fishes, the ostracoderms were largely bottom dwellers, though the members of one group, the Anaspida, which have a hypocercal tail fin, may well have fed on small organisms in the surface waters. Few of the placoderms look like real mid-water fishes, but with the evolution of the hydrodynamic coupling between hydroplane-like pectoral fins and a heterocercal tail fin in the cartilaginous fishes, more exploitation of mid-waters was possible. Even so, this exploitation was largely left to the bony fishes, particularly to the teleosts. The teleosts evolved a swimbladder with means of staying near or maintaining neutral buoyancy. By late Eocene times patterns of fish life were probably much as we see them today.

SURFACE FEEDERS

Many kinds of organisms live near the interface between air and water. Over rivers and lakes insects fly so close to the water that many become trapped in the surface film, which is also the natural support for the movements of other species. Adult, larval, and pupal insects move up to the surface to gather air. Spiders and insects fall on to the water from overhanging vegetation. Near the surface of lakes, particularly at night, move small planktonic animals. Even more is this true of the surface waters of the ocean. Indeed, in the upper few centimeters of sub-tropical and tropical regions there exists a special neustonic fauna. Most of these animals, belonging to such groups as the coelenterates, ctenophores, chaeto-gnaths, tunicates, copepods, mysids, stomatopods, deca-pods, cephalopods, and teleost fishes (young stages) are colored a deep-sea blue (Herring, 1967). Toward night-fall, many kinds of epipelagic and mesopelagic animals of the zooplankton move up and occupy the surface waters.

In evolving to exploit these surface aggregations of life, certain kinds of teleosts have independently ac-quired a number of similar features. In fresh waters, such fishes have arisen among the characoids, cyprinids, atherinids, and cyprinodontoids (toothed carps). Flying-fishes and half-beaks (Exocoetidae) and other atherinids are their counterparts in the sea (Figure 46).

Consider fresh waters and the most diverse of cyprino-dontoids, the top-minnows (Poeciliidae), which also occur in brackish and salt waters. Many species of this family have these features: (1) the jaws are small and usually upturned; (2) the premaxillae form a rounded fore border to the snout; (3) the snout, which is relatively wide, is flattened on top; (4) the origin of the dorsal fin is behind the mid-point of the standard length; (5) the lateral line system is reduced and/or modified; and (6) the body is usually fusiform and when analyzed by the Gregorian system of coordinates, av, the vertical distance between the pelvic origin and the longitudinal axis tends

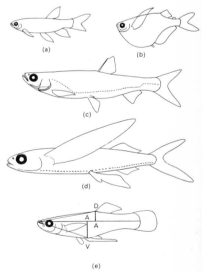

46 Surface-feeding fishes. (a) a characid, *Pyrrhulina nigrofasciata*; (b) a silver hatchet-fish, *Gasteropelecus maculatus*; (c) a cyprinid, *Cultriculus kneri*; (d) a flying-fish, *Prognichthys gibbifrons*; (e) a poeciliid, *Gambusia*, showing the relative position of the dorsal and pelvic fins and the greater length of AV compared to AD.

to be greater, and in advance of *ad*, the vertical distance between this axis and the dorsal origin (Figure 46; Greenway, 1965).

Not every top-minnow has all these features. *Belonesox* bears long, rapacious jaws, and in some species of *Xiphophorus* the body is deep and the dorsal fin originates near the mid-point of the standard length. But *Anableps*, which is placed in its own family, fits the above features very well, except that its jaws are not upturned. Indeed, it is so well adapted to a surface-dwelling existence that the eyes are divided horizontally into sections for aerial and aquatic vision. *Oryzias*, *Horaichthys*, the cyprinodonts, and goodeids also fit in all or most respects.

The small, upturned jaws of these fishes are neatly suited to the picking of food organisms on or near the surface. For instance, just before *Fundulus chrysotus* and *Xiphophorus helleri* seize food at the surface, the jaws are protruded as the mouth opens. Both species are equally adept at taking food from the bottom of an aquarium (Alexander, 1967a). When swimming and feeding at or near the surface, the body axis of these and other cyprinodontoid fishes makes an acute angle (usually less than 45°) with the water film. Clearly, these fishes, and most of them are small, will move most freely and accurately if their dorsal fin is not impeded by the surface film. The backward insertion of this fin will prevent such drag. But the mechanics of their staying at an acute angle to the surface have never been properly investigated. In the smaller species, at least, if the flattened snout is brushing the surface film the head will be held to some extent by surface tension. The flow of water over the rounded underside of the snout, which is usually wedge-shaped, will also tend to lift the head. Lift may also be obtained from the pectorals.

For certain cyprinodonts at least, there is a further advantage in not disturbing the surface film. *Aplocheilus lineatus* not only perceives surface waves, such as those made by a struggling insect, but is also able to find the source. The fish uses sets of modified supraorbital neuromasts, each formed in a groove and bearing a flag-like cupula (Figure 47). Surface waves are perceived only when they impinge on the narrow axis of the cupula. This fish gains an impression of the distance of its prey from the organs of one side alone, but both sides must work together to signal direction (Schwartz, 1965).

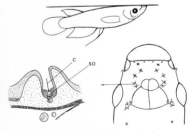

47 Above: a cyprinodont, *Aplocheilus lineatus,* shown just below the surface. Below right: dorsal surface of head, showing the position of the lateral-line organs, which are set in grooves. Below left: section through a lateral-line organ; *c*, cupula; *so*, sense organ. (Redrawn from Schwartz, 1965.)

Aspects of Convergent Evolution

Many other toothed carps have a head lateralis system much like that of *Aplocheilus* and we presume that their neuromasts function in somewhat similar fashion. Moreover, the visual powers of surface-seeking fishes are restricted. "Since surface fishes live very close to the water surface, their upward visual field (and hence the aerial window) is smaller than that in fish frequenting deeper water. Vision at the water surface (water-air interface) is practically impossible because of reflections and refractions of light. Yet surface feeders take food (prey) predominantly from the surface. These fishes seem to compensate for the dead band between the upward and downward visual field by means of their lateral line, which appears to be highly specialized" (Schwartz, 1967).

Certain characids are reminiscent of top minnows. Indeed, the genus *Grundulus* has been mistaken for one (Weitzman, 1962). Species of *Pyrrhulina, Copeina,* and *Nannostomus* tend to swim in the upper waters and frequently snatch food from the surface film. The related hatchet fishes (Gasteropelecidae) have similar habits and so do some of the cyprinids (for example, species of *Chela, Oxygaster, Esomus, Rasbora, Cultriculus,* and *Pelecus*). These fishes resemble the cyprinodontoids in the upturned gape, flattened bow-like snout, certain coordinates of form, and the posterior insertion of the dorsal fin. Little is known of their lateral-line function, but the reduction of the head system in *Pyrrhulina filamentosa* resembles the condition in certain cyprinodonts.

In the sea, silversides (Atherinidae), half-beaks, and flying-fishes (Exocoetidae) feed near the surface on planktonic animals. Nearly all these fishes share the features listed in the preceding paragraph, though most half-beaks have a bill-like jaw. Moreover, like the cyprinodontoids, the atherinids tend to lose the lateral line along the body. But in exocoetids—and other synentognaths—the lateral line is well developed and runs along the underparts of the body. In fact, the elaboration may be such that each neuromast can be stimulated through several tubules, which are directed downwards. The setting and openings of the lateral line thus indicate that the fish is warned of hydrodynamic displacements below its body. Moreover, the optical axes of synentognaths are canted downward slightly below the horizontal plane (Breder, 1932). When a flying-fish or half-beak is close to the surface it is particularly vulnerable to predators, though they may escape by taking to the air. During all or part of this flight the tip of the longer lobe of the caudal fin oscillates rapidly across the line of motion. But how does such a caudal work entirely under water, where its action would surely tend to cant the body downwards, which posture is hardly suitable for a feeding fish, although the upper caudal lobe is more flexible than the lower, both in the vertical[13] and horizontal planes? This is largely because the upper caudal rays have less overlap with the hypural base. Functionally then, the tail fin of a flying-fish may be hypocercal and thus tend to raise the head. But observations are needed.

Considering their vulnerability from below, the evolution of flight in certain surface-feeding fishes is not entirely remarkable. In fresh waters, *Pantodon* and the hatchet fishes have converged in developing pectorals that are vibrated rapidly in the latter by massive flight muscles. To base these muscles, the coracoid bones of hatchet fishes are expanded and used in a single fan-like structure. Except for this development and their wings, they would look like surface-feeding characins (Figure 46). A keeled thorax and large wing-like pectorals are also formed in the characid genus *Triportheus*, which looks very much like a surface feeder. Moreover, in these two respects species of *Triportheus* resemble the clupeids, *Raconda russelliana*, *Opisthopterus tartoor*, *Ilisha macrogaster*, *Odontognathus* spp., and the cyprinid genus *Chela*. After drawing attention to these facts. Weitzman (1960) remarks that it is an excellent instance of convergent evolution.

Most species of *Chela* school in the upper water layers, where they find much of their food. In this and other cyprinid genera with surface-feeding species (for example, *Esomus*, *Rasbora*, *Cultriculus*, *Nematobramis*, and *Luciosoma*) the lateral line runs well below the horizontal axis of the fish. This condition is similar to that of the flying-fishes and their relatives. The convergence is presumably related to the selective advantage of having a warning system against attack from below.

Surface-feeding fishes thus display convergences that are linked to the exploitation of their adaptive zone. But before the extent of these convergences can be assessed, one has to consider the relationships of the killifishes

13. Most of the interradial musculature is concentrated on the rays in the upper lobe.

(cyprinodontoids), silversides (atherinids), flying-fishes, and so forth, which Rosen (1964) believes form a natural group (Atheriniformes). Rosen envisages an ancestor of a basically hemirhamphid form with normal jaws. The resemblances between the above three groups could thus be due to common inheritance or parallel evolution. Even so, there still remain the striking convergences between atheriniforms and surface-feeding kinds of characoids and cyprinid fishes. Though the last two groups belong to the same order (Ostariophysi) the relevant resemblances between them are most likely to be convergent. Here, then, at the very least, are three groups that have independently given rise to surface-feeding forms. Curiously enough, such an adaptive type is rare among the great group of percomorph fishes. Notable exceptions are some of the gray mullets, which are equally adept at feeding at the surface or along the bottom. Gray mullets were once thought to be related to silversides, but their resemblances, which are largely in fin pattern and body form, are convergent. Moreover, in essential respects they have acquired independently a protrusile jaw mechanism like that of killifishes.

ADAPTIVE TYPES OF THE OPEN OCEAN

The radiations of fishes into the open ocean have produced five main adaptive types: epipelagic, mesopelagic, bathypelagic, benthopelagic, and benthic. The last four types are deep-sea fishes and some of the convergences between them are evident or implicit in discussion elsewhere.

Epipelagic fishes. Epipelagic fishes spend all or much of their lives in the euphotic zone, which reaches to a level of about 100 meters in the clearest oceanic waters. In the warmer parts of the ocean the lower limit of their occurrence is around the seasonal thermocline; here they swim in surface waters that are mixed to isothermal consistency by the trade winds.

Besides the flying-fishes and their allies, epipelagic fishes include various sharks, the opah (*Lampris regius*), ribbon fishes (*Regalecus and Trachypterus*, and so forth), carangids (for example, *Elegatis* and *Megalaspis*), bramids, dolphin fishes (*Coryphaena*), scombrids (tunas, and so forth), billfishes (istiophorids and *Xiphius*), and

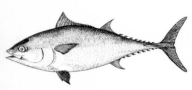

48　Convergence of form and fin pattern in a mackerel shark, *Lamna nasus* (above) and a blue-fin tuna, *Thunnus thynnus* (below).

ocean sunfishes (Molidae). Most of these are shapely, fusiform fishes, convergent in aspects of color, body form, and fin pattern. Here, though, we shall confine ourselves to the most specialized forms, which are the isurid sharks and the tuna-like scombrids.

The isurid sharks include the makos (*Isurus* spp.), mackerel sharks (*Lamna* spp.), and the great white shark (*Carcharodon carcharias*). All show far-reaching resemblances to the tunas (*Thunnus* spp.) and their allies (*Katsuwonus, Euthynnus, Auxis,* and *Allothunnus*). Convergences in outer appearance are: (1) the back is bluish or slate-colored and fades to paler underparts; (2) the body is close to an ideal streamlined form; (3) the first dorsal fin is set close behind the pectorals; (4) before the caudal fin there are one or more dorsal and anal finlets; (5) the caudal fin is sickle-shaped (with a high aspect ratio); and (6) each side of the narrow caudal peduncle bears a streamlined keel (Figure 48). Internally, isurid sharks and tunas converge in these respects: (1) the gills are large and have a great surface area;[14] (2) the myotomes, which form more than 60 percent of the total weight, have extensive forward and backward cones; (3) the red muscle of the myotomes is well developed; (4) the corpus of the cerebellus, as might be expected in such muscular fishes, is very large and elaborate; and (5) under the skin of the flanks is an elaborate plexus of blood vessels.

Most of these convergences reflect the life of these fishes. Isurid sharks and tunas are restless, far-ranging animals, capable of bursts of high speed (25 knots or more). Hence the laminar shape of the body, tapering to a slim but strong peducle with its lunate fin. Such a tail complex can be oscillated rapidly, and at least in the tunas, at high angles of attack (see Fierstine and Walters, 1968, who quote C. L. Hubbs' observations that mackerel sharks move the tail in a tuna-like way). The first dorsal fin, set above the center of gravity, and the caudal keels (Watts, 1961), may well enable these large and active fishes to turn quickly in pursuit of lively and elusive

14. In the tuna-like fishes (*Thunnus, Euthynnus,* and *Katsuwonus*) the secondary lamellae between adjacent gill filaments are fused together. These genera also converge with the wahoo (*Acanthocybium*) and the billfishes in developing fusions between the filaments. Muir and Kendall (1968) look on these modifications as associated with large gills and the "common habit of fast continuous swimming involving passive or 'ram' ventilation."

Aspects of Convergent Evolution

prey, such as squids and fishes. Massive, deeply nested muscle segments and especially their conspicuous red components, are also apt. Red muscles, as we are now aware, are built and fueled as the cruising motors of fishes.

In tunas the red muscle, and to a lesser extent the white muscle, are fed with blood from a subcutaneous counter-current system, which is formed from an artery and vein pair just above and below the horizontal septum. Retia mirabilia form around the red muscle and there are also branches to the white fibres (Figure 49). The essence of such vascular design is countercurrent interplay of arterial and venous blood. Indeed, this is why tunas are warm-blooded (Carey and Teal, 1966a and b). Oxygenated blood to the muscles, which reaches sea temperatures during its passage through the gills, meets outgoing blood warmed by muscular metabolism. Most of this heat (Carey and Teal's measurements of a blue-fin tuna indicate a retention of 98 percent in the red muscle) is transferred to the ingoing blood. In a sea temperature of 19.3°C, the highest temperature (31.4°C) was found in the red muscle of the blue-fin. Temperatures in the white muscle increased (from 21.3 to 29.3°C) from the surface to points near the backbone.

Carey and Teal discovered the same kind of subcutaneous, countercurrent blood system in a mako shark, the porbeagle, and the great white shark (Figure 49). One individual of the first species had a muscle temperature of 6°C more than the ambient level (20°C). But swordfish, dolphin fish, flying-fish, barracuda, silky shark, mackerel, Spanish mackerel, and rainbow runner, which are without such a subcutaneous circulation, had inner temperatures within a degree of sea temperatures.

Tuna-like fishes and isurid sharks have thus evolved closely convergent blood systems for keeping the temperature of their swimming muscles above that of the sea. There are, as Carey and Teal (1966b) point out, several advantages of being warm-blooded: "Digestion, the speed of nerve impulse transmission and metabolic reactions is in general speeded up. The rates of enzymatic reactions increase two to four-fold with each 10° rise in temperature. Muscle contraction and relaxation will be three times faster while the force of each contraction will remain the same. Thus, the fish can obtain three times as much power from the same mass of muscle by increasing

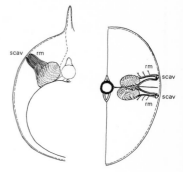

49 Transverse sections through the abdominal region of a mackerel shark, *Lamna nasus* (left) and the tail of a blue-fin tuna, *Thunnus thynnus* (right), showing the retia mirabilia (*rm*) that circulate blood to the red muscle (stippled).

its temperature 10°C. Tuna have been clocked at 70 kilometers per hour for a 10–20 second sprint, a greater speed than any other marine mammal. Speed is proportional to the ratio of power over drag. A look at a tuna shows that it is an extremely streamlined animal and it is unlikely that drag could be further reduced. Therefore, they probably owe their great speed to the extra power available from their warm muscles." A particularly significant feature is that these physiological advantages are most enhanced in the red muscles, the cruising motors of these superb pelagic fishes. It would be interesting to know how warm-blooded tuna and isurid sharks are when they migrate into temperate waters. There is one structural indication. In the tuna-like fishes, the subcutaneous vascular plexus is best developed in the genus *Thunnus*, the species of which make the most wide-ranging migrations. In fact, the blue-fin tuna, as Carey and Teal (1969) found, is able to control the temperature of its body between 25–30°C so that the warmest part of the muscle varies only by 5°C over a 10–30°C range of ambient water temperatures. Such powers of thermoregulation give the blue-fin tuna the freedom to range from Norwegian to Bahamian waters, where temperatures go from 6–30°C.

Tuna-like fishes and isurid sharks are surely the most highly developed species of the epipelagic zone. Their convergent features give them the means to cruise tirelessly and, if need be, to move swiftly in the sea. To keep their highly organized bodies in motion, it is no wonder that the most specialized forms have evolved the means to seek the most productive waters of the tropical and temperate ocean.

The mesopelagic zone. Convergences in the adaptive features of mesopelagic fishes are evident or implicit in earlier discussion. Here, then, we need only review briefly the relevant facts. For convenience, I repeat that the main groups of mesopelagic fishes are isospondylous (searsids, alepocephalids, stomiatoids, argentinoids) iniomous (alepisauroids and myctophids), melamphaid chiasmodontid, and trichiuroid. The populations of these fishes are centered at levels between 150 and 1,000 meters, which distance more or less spans the twilight zone of the ocean. Convergences between mesopelagic fishes involve their coloration, light organs, sense organs means of making a living, and life history patterns.

Aspects of Convergent Evolution

Dark backs and silvery sides have been acquired independently by some of the stomiatoids (for example, certain gonostomatids and sternoptychids), argentinoids (*Opisthoproctus*), and various myctophids. The concealing value of such a color pattern in the twilight zone, where at a given level the downward component of the light field exceeds the lateral and upward components, has already been disucssed. Instead of holding vanishing mirrors, other fishes have evolved fine mosaics of dermal melanophores, which reflect very little light. Such convergent cloaks are found on predatory stomiatoids, melamphaids, chiasmodontids, and trichiuroids.

Certain searsids, alepocephalids, argentinoids, paralepidids, and all of the myctophids have light organs. Except perhaps for those of the first two groups these organs were independently acquired. The simplest organs may be no more than a luminous gland cowled by black skin; in the most elaborate kinds the cowl is faced with a silvery reflecting layer and there is a lens to concentrate the light. This is close in spectral range to that of mesopelagic sunlight. And it is hardly due to chance that these blue-green rays are also the most readily absorbed by the visual cells of mesopelagic fishes.

Each group of fishes has evolved distinct patterns of lights and the more elaborate ones have their own design. The most striking convergence is that some of the more complex organs or the entire system in opisthoproctids and paralepidids cast their light downwards. The presumed advantage of this in obliterating the ventral silhouette is considered in Chapter 3. Individual patterns of light may serve as recognition signs between members of a particular species. Moreover, each sex may bear its own luminescent signs. The system of caudal glands in myctophids, the male usually developing them on top of the peduncle, the female below—is unique to this group. In addition, the sexes of *Diaphus lewisi* are distinguished by the size of the antorbital lights, which are much bigger in males (Nafpaktitis, 1966). This difference is also true of certain stomiatoids (for example, *Idiacanthus, Aristostomias, Photostomias*), but here the males have larger cheek lights than the females.

Life in the twilight zone has evoked elaborate convergences between the eyes of mesopelagic fishes. The advantage of evolving another very sensitive system for locating prey, predators, and members of the same species is related to convergences in the design of the lateral-line organs. For instance, searsids, myctophids, evermannellids, and melamphaids not only have large and complex head neuromasts, which are set in wide canals, but also free neuromasts on the head. In the free organs, there is a curious convergence between their design and disposition on the head of *Melanonus* spp., *Pseudoscopelus* spp., and *Odontostomops*. Most of the neuromasts are unusual in being elliptical in outline and their functional axis is presumably the longer one. The organs are so disposed that their axes tend to be aligned along the horizontal axis of the head. It would be interesting to know the functional significance of these convergences.

Many mesopelagic fishes, represented by stomiatoids, myctophids, paralepidids, and trichiuroids, migrate diurnally to and from the surface layer (Marshall, 1960). The biological significance of these movements, which are probably trophic, is discussed in Chapter 3, but have these groups acquired this habit independently? Circadian rhythms are certainly widespread in plants and animals and even in single cells. All one can say at present is that if mesopelagic fishes have endogenous rhythms, these are easily overriden by unusual changes in the light regime. For instance, deep scattering layers, which contain mesopelagic fishes, move upwards during a solar eclipse. Moreover, some species may not migrate at all. Presumably the migrators, which have highly sensitive eyes, are able to detect the approach of sunset and sunrise, the times of their upward and downward migrations. But are their diurnal rhythms simply the outcome of keeping to particular isolumes? Until we can keep these fishes under controlled conditions, the question of endogenous rhythms will remain unsettled.

Species that feed on zooplankton, such as searsids, myctophids, and melamphaids, converge in developing rows of small gripping teeth along their jaws and a relatively small pyloric part of the stomach. This part of the gut is much extended in fishes living on larger prey, such as the predatory stomiatoids, alepisauroids, chiasmodontids, and trichiuroids. These forms also bear long stabbing teeth in the jaws, palate, and pharynx, and the gill rakers are usually reduced to tooth-like structures. Members of the first three groups have also converged in acquiring a very distensible stomach and body wall, which enable them to contain prey that may even be larger than themselves. Presumably, these distensible parts contain an elaborate system of elastic fibres.

Except perhaps for certain searsids and alepocephalids, mesopelagic fishes spend their larval life in the productive, surface layer of the ocean. The eggs are likely to be shed in the depths and develop as they rise towards the surface. During and after metamorphosis the young move down to the adult living space. If the production of eggs is enough to counter the heavy mortality of eggs and larvae—and this must usually happen—the advantage of the above life history pattern is that the young are dispersed over the living space, which may be immense. Indeed, they may even be carried to areas where vegetative, but not reproductive, existence is possible. But there is always the chance, as O'Day and Nafpaktitis (1967) point out, for genetic changes to convert vegetative into reproductive individuals. Thus the life history patterns of mesopelagic fishes enable them to exploit, and even explore, wide stretches of the ocean.

A life history depends on a complex sequence of events. Has the pattern outlined above, which is also shared by bathypelagic teleosts, been evolved independently by each major group of mesopelagic fishes? Their common ancestor, perhaps a fish rather like the holostean *Leptolepis*, may have reproduced by pelagic eggs and larvae. If this ancestor, as is most likely, lived in shallow seas, changes would be needed, for instance, in mating behavior and egg structure, to adapt the descendents to a mesopelagic existence. Convergences rather than parallelisms would arise, for it seems unlikely that the genetic constitution of the ancestor would have predisposed diverse descendants to the same "deep-sea" mutations. Most likely, the life history patterns of mesopelagic fishes are convergent simply because this is the only way of providing the larvae with a proper nursery ground. At all events, no primary mesopelagic teleost has become viviparous.

The bathypelagic zone. The most diverse bathypelagic fishes are ceratioid anglers (about 100 spp.); the most numerous individuals belong to the black species of *Cyclothone. Gonostoma bathyphilum*, related to *Cyclothone*, forms part of this fauna, and so do the gulper eels (Lyomeri), bob-tailed snipe eel (*Cyema*), certain cetomimids, and a few species of macrourids and brotulids.

This fauna lives in the water masses below a depth of 1,000 meters. Apart from *Cyclothone* and *Gonostoma*, these fishes are distant relatives and their many similarities, now to be considered, must surely be convergent.

Except for the brotulids (pallid) and the cetomimids (red), bathypelagic fishes are black or brown. Red and black skins reflect very little luminescence (Nicol, 1958) and but for this illumination, one presumes that the fishes of this sunless environment, like cave-dwelling species, would have lost most of their pigmentation.

Apart from the free-living males of ceratioid anglers, the eyes are small or regressed. The lives of these forms that keep their eyes are presumably dependent on the recognition of luminescent events. In ceratioids, gulper eels, and *Cyema* there are no lateral-line canals; all the neuromasts are borne on papillae, stalks, or flaps. How this condition arose may be demonstrated by the lateralis system on the tail of the two whale fishes, *Cetomimus craneae* and *Ditropichthys storeri*. Both species have wide canals opening through large pores, which in *Ditropichthys* are so large as to take up more space than the covered parts of the tail canal (Figure 50). In the other species, where the covered sections are more spacious than the pores, each neuromast rests on a diamond-shaped pad of connective tissue. The neuromasts of *Ditropichthys* are set at the end of wide flaps of connective tissue, which is like the entire lateralis system of certain ceratioids. Thus, stalked neuromasts may have arisen as the canals disappeared.

The convergent evolution of stalked neuromasts by bathypelagic fishes is probably related to their slowly moving environment and their stealthy ways of making a living. In part, the same factors may well be behind the evolution of large olfactory organs on the males of most ceratioid anglers, *Gonostoma bathyphilum*, and *Cyclothone* spp.

These macrosmatic males are smaller than the females, which may attract their partners through the production of specific pheromones. The males are well equipped to hunt for a mate, for their myotomes contain a well developed outer layer of red fibres, which have inbuilt stamina. The waiting females have very little red muscle. Males mature much earlier than the females and easily outnumber their ripe partners. The evolution of small or

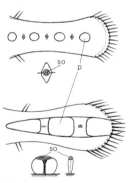

50   The tail region of two whale-fishes. Above: *Cetomimus craneae*. Below: *Ditropichthys storeri*. Note the much larger canal pores (*p*) of the second species. Each sense organ (*so*) of the first is set on a diamond-shaped pad of connective tissue, while each organ of the second is carried at the end of a broad flap of skin.

dwarf males must also reduce the competition for food between the sexes and reduce the energy budget of a species.

The energy requirements of all bathypelagic fishes are greatly reduced by stringent economies of their tissue systems. They are thus able to exist in waters that hold very low standing stocks of potential energy.

The benthopelagic zone. Observers in underwater vehicles and many photographs have now established firmly that diverse fishes habitually swim near the deep-sea floor. Most of these fishes; synaphobranchid eels, halosaurs, notacanths, macrourids, morids, and brotulids, contain a well-developed swimbladder. The others, alepocephalids, ateleopids, chimaeroids, and certain squaloid sharks (for example *Oxynotus*, *Centroscymnus*, and *Echinorhinus*) are without a swimbladder. Even so, some squaloids, including *Centroscymnus coelolepis*, have a very large liver filled with the hydrocarbon squalene. The low specific gravity of squalene (0.86) and its high concentration (up to 90 percent) in a voluminous liver (about 25 percent of the total volume) brings these sharks close to neutral buoyancy (Denton, 1963). Ateleopids and alepocephalids may well be close to neutral buoyancy largely because of their flimsy skeleton and weak myotomes (see Denton and Marshall, 1958), and the chimaeroids probably derive most of their uplift from a large, oily liver and their mobile pectoral fins. Members of the benthopelagic fauna thus exert little or no energy to stay at a particular level as they move over the oozes. They are much freer than the earth-bound benthic species to feed on the crustaceans and other organisms that swim over the deep-sea floor. They are also free to hover over the oozes and probe them for food.

Benthopelagic fishes are most diverse over the continental slopes. Indeed, the chimaeroids are virtually confined to the slope. The slope dwellers have large eyes with exaggerated dioptric parts and highly sensitve retinae. Halosaurs, macrourids, morids, and brotulids also converge in developing wide lateral-line canals on the head, which contain large neuromasts. There are also free-ending organs on the head and sometimes on the body. Such lateralis systems are likely to be versatile and sensitive, helping their owners to keep in touch and find their food in dim or dark surroundings.

Benthopelagic fishes converge in two respects with regard to the swimbladder. The deeper the living space, the longer are the retial capillaries that feed the gas gland (with increasing depth there is increasing need to maintain the efficiency of the gas-conserving and gas-secreting function of the retia). Second, the two most diverse groups, the macrourids and brotulids, converge in their sonic dimorphism. In many species the males alone develop drumming muscles on the swimbladder. However, the design of the drumming mechanism differs considerably in the two groups (Figure 26).

Far-reaching convergences between benthopelagic fishes concern their form, fin pattern, and jaws. Ateleopids, halosaurs, notacanths, and most of the macrourids have the jaws below a projecting snout. There is also a long tapering tail, fringed by a long, many-rayed anal fin. The pectoral fins are set laterally, but in halosaurs, ateleopids, and macrourids there is a short-based dorsal fin (Figure 51). Such a fin pattern is correlated with the underslung jaws. In particular the undulations of the anal fin and tail will tend to depress the head, so bringing the jaws into the right position for taking food from the sea floor.[15] Indeed, macrourids and halosaurs root in the oozes (Marshall and Bourne, 1964). Both groups converge further in developing spectacles over the eyes.

These convergent groups of long-tailed fishes comprise well over half of the benthopelagic fauna. Moreover, some of the brotulids and morids have a slender tail, though the anal fin is not differentially enlarged. Indeed, most of the benthopelagic fishes that live at depths greater than 3,000 meters are "rat-tails." Apart from the correlation between fin pattern and jaw mechanism, why have "rat-tails" succeeded so well in the benthopelagic zone of the deep-sea? An elongated tail can, of course, carry a long-based lateral line, which gives extra means of detecting prey and predators (Orkin, quoted by Wynne-Edwards, 1962). Lastly, fishes with many-rayed median fins and a tapering tail are well equipped to undulate steadily and slowly over the oozes. If they pass over food or detect it behind the head they can soon reverse direction and snap it up. The same is true for the eels.

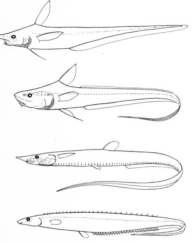

51 "Rat-tailed" benthopelagic fishes of the deep sea. From top to bottom: *Ateleopus indicus*; *Lionurus filicauda*, a macrourid; *Aldrovandia*, a halosaur; *Macdonaldia rostrata*, a notacanth.

15. Notacanths snip off pieces of bryozoans and sea anemones.

Aspects of Convergent Evolution

The benthic zone of the deep sea. In polar and temperate regions the most successful groups of benthic fishes are zoarcids and liparids. The bathypteroids, ipnopids, and chlorophthalmids, which are related groups, are better adapted for subtropical and tropical regions. These fishes, which are without a swimbladder, spend most of their life resting on the deep-sea floor. There are few evident convergences, though a particularly striking one concerns reproductive adaptations in bathypteroids (including ipnopids) and in chlorophthalmids. Ovotestes are developed in both groups.

Benthic fishes of coastal seas. When they are not migrating, feeding, courting, mating, and so forth, the habit of benthic fishes is to rest quietly on the bottom. Most of the rays (Batoidei), certain sharks, notably angel fishes (Squatinidae) and carpet sharks (Orectolobidae), and a great many teleosts have such habits. Here we shall be concerned largely with the teleosts. We think of groups such as the weevers (Trachinidae), stargazers (Uranoscopidae), dragonets (Callionymidae), nototheniiforms, blennioids, eel pouts (Zoarcidae), sea snails (Liparidae), flatheads (Platycephalidae), bullheads (Cottidae), sea poachers (Agonidae), lumpsuckers (Cyclopteridae), flatfishes (Heterosomata), toadfishes (Haplodoci), clingfishes (Xenopterygii), anglerfishes (Lophiidae), and batfishes (Ogcocephalidae). (See Figure 52.) Nearly all these fishes have lost the swimbladder, though it may be present in the larvae of a few species. The exceptions are the toadfishes and the petroscirtine blennies. But the swimbladder of toadfishes is small, much less than 5 percent of the total volume of the fish, the figure required for a marine fish to be neutrally buoyant. In fact, toadfishes use their swimbladder, which is provided with drumming muscles, as a sound maker. The petroscirtine blennies (except *Xiphasia*) have a well-developed swimbladder and are easily able to swim at mid-water levels.[16] They are not true benthic species, which are heavier than their environment.

A striking convergence displayed by these fishes is in their color pattern, which is not often plain but generally broken into a cryptic pattern. Weevers and stargazers bury themselves in the sediments so that only the forepart of the head emerges, while flatheads and many

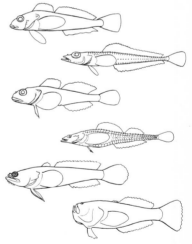

52 Benthic fishes. From top to bottom, *Bovichthys diacanthus* and *Prionodraco evansi* (nototheniiform fishes of the Southern Ocean); *Artediellus europaeus*, a cottid; *Occa dodecaedron*, an agonid (Arctic seas); *Chatrabus damaranus*, a toad fish; *Ichthyoscopus lebeck*, a stargazer. Note the convergence of form and fin pattern in the two upper pairs of fishes.

---

16. Many of the gobies, which will be considered later, also have a swimbladder.

flatfishes cover themselves with a fine layer of sand or mud. Whatever the device for concealing themselves, most benthic fishes lie in wait for their prey and seize it suddenly. Notable exceptions are the flatfishes, many of which may cover much ground in search of food.

Some of the flatfishes and other flattened fishes, rays and flatheads, and the stargazers, converge in that their eyes bulge well above the surface of the head. Moreover, representatives of all four groups have evolved opercular shades on the eyes. The operculum is formed along the upper edge of the pupil and in bright light it expands so that the pupil is almost occluded (Figure 53). These fishes are thus not "dazzled and defenceless' as Walls (1942) says, when exposed to the sun, particularly over shallow bottoms with a high albedo.

Benthic teleosts converge also in their means of ventilating the gills. They have relatively large gill chambers and expansive gill membranes. Experiments on a cottid, a dragonet, and two flatfishes have shown that the suction-pump action of the gill chambers is much more effective than the pressure pumping of the buccal cavity for drawing water over the gills. Similarly, in a ray the power of the parabranchial suction pumps was found to be greater than that of the buccal cavity (Hughes, 1960). Accentuation of the suction mechanism is apt in fishes that spend much of their time at rest on the bottom. As Hughes concluded: "Selection has favoured the evolution of a mechanism which is well adapted to drawing a current across the gills during a long part of the respiratory cycle. Such mechanisms are admirably adapted to ensure a steady flow across the gills without creating any disturbance of the muddy or sandy bottom." Prey and predators must not be alerted, for the life of these fishes, as we saw, is centered on concealment.

Compared to active mid-water fishes, which have a large gill area per unit of body weight, benthic teleosts have a smaller respiratory surface. For instance, contrast the mackerel (*Scomber scombrus*), with over 1,000 mm^2 of gill surface per gram of body weight, to a toadfish (*Opsanus tau*), which has no more than about a fifth of the surface (see also Gray, 1954). Moreover, active mid-water fishes not only have a large gill area formed from a large number of lamellae, but also a very thin gill epithelium separating blood from water. In contrast, not only are the gills of benthic fishes coarser,

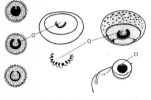

53   Opercula (*o*) on the eyes of fishes. Left: from top to bottom, three stages in the expansion of the operculum of the loricariid catfish, *Plecostomus*. Middle: the scallped operculum of a skate, *Raja clavata*. Top right: the eye of a brill *Scophthalmus rhombus*. Bottom right: the eye of a stargazer, *Uranoscopus scaber*. (Redrawn from Walls, 1942.)

but the diffusion distance between water and blood is considerably greater (Berg and Steen, 1966).[17] Considering the mode of life of benthic teleosts, these differences and convergences are understandable.

Except for some of the flatfishes, benthic teleosts have relatively small axial muscles. Moreover, the caudal fin has a low aspect ratio and generally a reduced number of principal rays. Oscillations of the tail are used only when these fishes are hard pressed. Cruising movements, or even dashes to secure prey, are effected by the large, fan-like pectorals in weevers, nototheniiform fishes, blennies, gobies, bullheads, and sea poachers. The muscles of these fins, unlike the axial muscles, are well provided with red fibres, which give a fish stamina for cruising.

Benthic fishes depend so much on camouflage, cover, and stillness that they have virtually lost one part of the nervous system. In a weever, a goby, two blennioids, and a sole, De Angelis (1950) found a reduced Mauthnerian system, which when fully formed is the basis of quick escape reactions. It should also be noted that some forms are well armed, particularly weevers, stargazers, dragonets, flatheads, bullheads, sea poachers, toadfishes, and batfishes.

There are also certain less pervasive convergences. The development of suction-cup-like pelvic fins, found in liparids, cyclopterids, gobies, and clingfishes, is the most remarkable. Even so, the first two groups are sometimes placed in the same family, which might mean that their pelvic fins were inherited from a common ancestor or as the result of parallel evolution. Freihofer (1963), who finds a similar pattern in the ramus lateralis accessorius nerves of the liparids and clingfishes, is inclined to think that their modified pelvics are not convergent. At all events, there is no doubt that the pelvic fins of gobies are suctorially convergent with those of the other groups. Moreover, when we consider the searching power of waves and tides, we can understand the independent evolution of ventral suckers to keep a fish clinging to stones, rocks, and plants. If "Bernoullis' effect" can make pebbles jump from the bottom, fishes not much denser than the sea need a good holdfast. It is hardly surprising

---

17. However, two "bloodless" chaenichthyids of the Southern Ocean have three or more times the relative gill surface of the sea scorpion, *Myoxocephalus scorpius*. These are *Chaenocephalus aceratus* and *Champsocephalus esox*.

that the disc of the liparids is most highly developed in tide-pool species. But the clingfishes, whose shape is convergent with that of the hill-stream homalopterids, are the most highly evolved. About New Zealand species, Morton and Miller (1968) write: "The little fish attach immovably to small stones as they come to rest, or when the sea breaks over their pools."

Lastly, toadfishes, clingfishes, cottids, cyclopterids, certain zoarcids, and most probably the nototheniids converge in reproductive adaptations. They produce relatively large, opaque eggs that are heavier than sea water and are fastened to some convenient surface near or on the bottom. With regard to the adaptive nature of these convergences, I wrote elsewhere: "In coastal waters the tides are usually higher than those in the open ocean. Waves search and scour the shallows, keeping in motion long-shore currents and rip currents. Anchored, non-buoyant eggs are suited to such surroundings, but floating eggs, aside from other hazards, would often be cast ashore. Yet an attached egg mass makes excellent food. Perhaps it is primarily to thwart egg-eaters that some species have taken to guarding their eggs, which habit is common among blennies, gobies, lump-suckers, clingfishes, bullheads and toadfishes. The protector is nearly always the male, who threatens intruders by fearsome displays and advances, even by grunting if he has the means. Some gobies and clingfishes also fan water over the eggs, which will keep them aerated and free from a coating of silt.

"The larvae that hatch from non-buoyant eggs are generally more advanced in development than those that emerge from floating eggs. In a good many species the newly-born larvae bear a small supply of yolk and are nearly ready to find their own food. The yolk-sac stage, when mortality may be severe, is passed within the egg" (Marshall, 1965, pp. 273–274).

Benthic freshwater fishes. Limnologists contrast standing-water (lentic) habitats with those that flow (lotic). Fast flowing lotic habitats remind us in some ways of intertidal waters, and we have already noted the convergence of form between the homalopterids of Asiatic hill-streams and the clingfishes. In addition to their flattened bodies, homalopterids, converge with clingfishes in developing a ventral sucker, but in the former the pectoral fins, as well as the pelvics, form part of the suction disc. In both groups the sucker is scaleless, and so is the entire body of clingfishes. Moreover, both homalopterids and clingfishes have the habit of making short darts from place to place, driven by rapid flicks of the tail. And they keep close to the bottom, where current speeds are reduced by frictional forces.

In the deepest lentic environment, Lake Baikal, bullheads (Comephoridae and Cottocomephoridae) dominate the benthic fish fauns. But the most diverse benthic forms, whether of lotic or lentic environments, are catfishes.

Not every catfish is benthic. Some of the Siluridae, Schilbeidae, and Mochokidae are mid-water forms (Alexander, 1965). But as Alexander notes, most catfish, like many other benthic fishes, are rather depressed in shape. They differ, however, from benthic fishes of marine environments in retaining a swimbladder. Even so, this is reduced below the volume needed (about 7 percent of the body volume) for neutral buoyancy in fresh waters. In a number of groups, such as the clariids, callichthyids, and loricariids, the swimbladder is essentially reduced to the forward chamber, which is encapsulated in bone. In this respect they converge with the homalopterids and loaches, and probably with each other. The functional significance of this modification is that the sensitive linkage between the ears and the swimbladder is retained. Moreover, in quickly moving waters, fishes with a relatively high density will be subject to frictional forces that will tend to keep them in position on the bottom (Alexander, 1965). But benthic freshwater fishes belonging to groups other than the Ostariophysi are like their marine counterparts in losing or virtually losing the swimbladder. The darters of North American freshwaters are good examples.

Catfishes remind us of certain benthic forms of coastal seas in at least three respects. One is the rather depressed form and has been noted above. In particular, *Chaca chaca* has a broad anglerfish-like head but this catfish angles with its barbels, not with a modified dorsal finray. Second, many catfishes develop large spines in their dorsal and pectoral fins, which must sometimes be effective against predators. Benthic marine fishes also develop spiny armament. As in catfishes, spines may well be a compensation for their relatively sedentary mode of life. Last, at least one catfish (the loricariid *Plecostomus*) has evolved a pupillary operculum, which can expand and so shield the eye from dazzle (Figure 53). The presence of such an ocular structure in skates, flatfishes, flatheads, and so forth, is considered elsewhere in this chapter.

General considerations

Convergence was first defined here as the independent acquisition of similar characters in lineages that have evolved in the face of similar environmental conditions. Convergent characters were contrasted with those due to inherited genetic potentialities (parallelisms), but virtually equated with analogous characters (similarities in overall function that are also independent of common ancestry). Even so, some convergences are not so much related to sets of environmental conditions or functions as to the general properties of the aqueous medium. This wider aspect of convergence will now be considered.

Fish-like forms were inevitable as active organisms evolved in water. Gardner (1964) has expressed this very well: "In the sea, for example, the ability to move about rapidly in search of food gave an animal great competitive advantage over sessile and slow-moving forms. A mouth is obviously more efficient on the front end of a fish than on its back end; the fish can swim directly towards food and gobble it up before some other animal gets it. This single feature alone, the mouth, is sufficient to distinguish the front end from the back of a fish . . . Other features, such as eyes, are also clearly more efficient at the front end, near the mouth, than at the back. A fish wants to see where it is going, not where it has been." It is also inevitable that ears with semicircular canals, which register angular accelerations, should be placed near the junction between the head and trunk, the most stable point of a swimming fish. It seems reasonable, too, that the olfactory organs should be placed at the front of the head. Apart from their unobstructed position, the nostrils will thus have the widest possible lateral sweep as the head swings from side to side. The chances of finding a scent trail and its source will thus be enhanced.

To resume the quotation from Gardner: "At the same time that locomotion was leading to distinction between front and back, the force of gravity was causing similar differences between an animals' top and bottom." But "there is nothing in the seas' watery environment to make a distinction between right and left significant. A swimming fish encounters a marked difference between forwards and backwards because one is the direction it goes, the other is the direction it comes from. If it swims up it reaches the surface of the sea. If it swims down, it reaches the ocean floor. But what difference does it encounter if it turns left or right? None. If it turns left, it finds the sea, and the things in it, exactly like the sea that it finds if it turns right. There are no forces, like the force of gravity, that operate horizontally in one direction only. It is for these reasons that various features—fins, eyes, and so on—tended to develop equally on left and right sides. Had there been a great advantage for a swimming fish to see only to the right and not the left, no doubt fish would have developed only a single eye on the right. But there is no such advantage. It is easy to understand why a single plane of symmetry remained dividing fish bilaterally into mirror-image right and left sides."

In view of such inevitability, did fishes (jawless and jawed) evolve from more than one stock of active protochordates. Judging from the neurosensory design of modern forms, a single origin is more likely. The central nervous system, the acoustico-lateralis system, the Mauthnerian system, the pituitary complex, and the caudal neurosecretory organs are so essentially alike in both jawless and jawed fishes that one is led to think of a common ancestor. There are also the Calcichordata, which have a fish-like brain with optic, hypophysial, medullary, and olfactory parts (Jefferies, 1968). Jefferies considers this subphylum, which appears to have certain affinities to the echinoderms, to have given rise to the chordates.

Fusiform bodies are another inevitable product of lively motion in water. We see this not only in fishes but also in ichthyosaurs, penguins, seals, and cetaceans. Among the fishes, the most adept swimmers have a body shaped very like the theoretical laminar form of the hydrodynamicist. Dolphins and porpoises also have a laminar shape, and certain of the former, at least, have evidently evolved a skin to counter a major hindrance to quick movement in water—turbulence. Pressures associated with turbulence are readily transmitted through an elastic and thin outer layer to an elaborate, oil-filled substratum, which is believed to absorb the small shocks of turbulence (Gérardin, 1968). Bill-fishes (Istiophoridae) may well have a similar kind of skin. The skin of tunas is most supple and oily behind the corselet, and it is here, beyond the point of maximum cross-section, that turbulence is likely to arise. But close study is needed. If we found that some fishes have "copied" the turbulence-damping skin of dolphins, it would not be so surprising. We might also recall the convergent development of "adipose eyelids" to reduce turbulence behind the bulging eyeballs of diverse fishes.

The convergent evolution of escape-mechanisms in active aquatic animals has been discussed. The development of giant nerve fibres in fishes, squid, and decapod crustaceans—fibres designed to quickly activate powerful propulsive units—was seen to be related to means of making a rapid getaway in an inert medium. This aspect needs no more elaboration, but there is another side to the inertia of water. Once water is set in motion the disturbance persists for a considerable time. It is surely this aqueous attribute that favored the evolution of sensory elements able to detect hydrodynamic displacements, whether caused by prey, predators, or social partners, and so forth. In essence, the lateral line of fishes is such a system of detectors. Each unit (neuromast) consists of a gelatinous cupula resting on hair cells and supporting cells. The hairs of a sensory cell consist of a kinocilium and numerous stereocilia, which range in number from 40 to 50 in *Lota*, the burbot (Flock, 1965). Adjacent hair cells are oriented with their kinocilia pointing in opposite directions, alternately toward the head and the tail, but always along the axis of the lateral-line canal. Each hair cell has a directional sensitivity to shearing movements of the cupula, which is directly or indirectly (if the neuromast is in a canal) disturbed by displacements of the medium. In the burbot, Flock found the third supratemporal neuromast to have a figure-of-eight polar diagram of sensitivity, the line between the two most sensitive coordinates (0 and 180°) coinciding with the axis of the canal.

Thinking now of the evolution of neurosensory systems in an inert liquid, it would be surprising if fishes (and some amphibians) were the only aquatic animals with displacement receptors. Cephalopods, surprisingly, do not appear to have a "lateralis" system. But arrow worms (Chaetognatha) have a well-developed series of free-ending organs (Horridge and Boulton, 1967). As in fishes, the sensory cells bear hairs or bristles (Figure 54). In *Spadella cephaloptera*, it seems cells with nonmotile cilia respond to nearby vibrations, and thus the animal is able to seize prey with great rapidity and accuracy. Pelagic nemertean worms also have free-ending "lateral-line" organs.

Hair cells develop also on the "fingers" of the ctenophore, *Leucothoe*, while the hydromedusan, *Eutonina indicans*, evidently has vibration receptors, for it will bend its manubrium toward a nearby source of vibrations. Horridge (1966) also draws attention to the lobster, *Homarus vulgaris*, which has bristle-bearing vibration receptors in pits over the anterior part of its body. Moreover, he suspects that the tentacle tips of tubicolous animals and the apical tufts of cilia on trochophore larvae, which have a nerve supply, are sensitive to vibrations. Such then, is the evidence for the widespread convergent evolution of displacement receptors in aquatic animals.

Last, there is the inherent factor of hydrostatic pressure, which increases steadily by 1 atmosphere for each descent of 10 meters. Though we can only surmise that physiological adaptations to high pressures have evolved independently in remotely related groups of deep-sea fishes, another aspects is quite evident. In the teleosts, control of specific gravity through a swimbladder, equipped with gas-secreting and resorbent systems, is the only solution to proper maintenance of neutral buoyancy. Such a closed swimbladder as described above has been acquired convergently more than once.

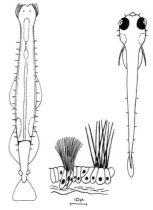

54   Comparison of the "lateral-line" organs of an arrow-worm (left) and a larval fish, *Oryzias* (right). (After Iwai, 1967.) Between the two animals is shown a tuft of setae and a fan of ciliated neurones from an arrow-worm, *Spadella*. (Redrawn from Horridge, 1966.)

ORGANIC CONSTRAINTS

In evolving, then, aquatic organisms are faced with constraints imposed by the physical nature of their environment. Convergences are thus bound to arise. There are also constraints inherent in the nature and number of parts that are available for a particular functional end. Thinking especially of neuroendocrine systems, Scharrer (1959) wrote thus: "Given the limitations of available materials and the basic principles of chemistry and physics one may expect that even very different types of animals arrive at analogous solutions of the same basic problems." More generally, Pantin (1951, 1965) has considered organic constraints in a simple but revealing way. In the earlier paper, he argued: "The organism is thus built up of standard parts with unique properties. The older conceptions of evolutionary morphology stressed the graded adaptations of which the organism is capable, just as putty can be moulded to any desired shape. But the matters we have discussed lead us rather to consider the organism as more like a model made from a child's engineering constructional set: a set consisting of standard parts with unique properties, of strips, plates and wheels which can be utilized for various functional objectives, such as cranes and locomotives. Models made from such a set can in certain respects show graded adaptability, when the form of the model depends on a statistically large number of parts. But they also show certain severe limitations dependent on the restricted properties of the standard parts of the set."

Blood vessels can form a "statistically large number of parts" in the form of capillaries. But there is only one way, by the close association in parallel of arterial and venous capillaries, that a countercurrent system can be formed. Thus, it is not surprising that retia mirabilia, which retain and develop gas tensions within the swim-bladder have evolved independently more than once. This system is capable of graded adaptation—through an increase in the length of the capillaries—to increased depth of living spaces. A countercurrent system can also be adapted to more than one end. In the tunas and isurid sharks, such a system has evolved convergently for the conservation of heat in the swimming muscles.

Lastly, there is a countercurrent circulation in the choroid layer of the eyes of *Amia* and diverse teleosts. Two eels (*Anguilla rostrata* and *Conger oceanica*) and three elasmobranchs (*Dasyatis centroura*, *Raja ocellata*, and *Mustelis canis*) are without a choroidal rete, while it is minute in certain bottom dwellers (for example, *Lophius* and *Opsanus*). Besides presenting these data, Wittenberg and Wittenberg (1962), correlate high choroidal oxygen tensions with well-developed retia, as in the cod, *Echeneis*, bluefish, and so forth. An elaborate rete is evidently needed to build high oxygen tensions, particularly for the retina. The presence of a choroidal rete in *Amia* and various teleosts is surely a further instance of convergent evolution. If convergence is not involved, we must look well down among the holostean fishes for a common ancestor. At all events, the other living holostean (*Lepisosteus*) is without a choroidal rete (Walls, 1942), and it certainly depends on its eyes to catch its prey.

At the other extreme, an outstanding instance of a mechanism involving a small number of parts is the jaw system of teleost fishes. The main outer elements are the premaxillae, maxillae, and dentaries. In the most primitive (isospondylous) forms the first two bones, which bear teeth, border the upper jaw, the head of each maxilla being hinged to the palatine bone. This kind of jaw has become more mobile in certain distinct ways. In the chanoid fish, *Phractolaemus*, the maxillae and dentaries form the four sides of a rhombus, which pivots on the quadrate bones at its posterior angle. The protrusion and retraction of the jaws is effected by this pivot. All other kinds of protractile jaws depend on a change of emphasis in the upper jawbones, whereby the premaxillae bar the maxillae from the gape. First, though, nonprotractile, maxillae-excluded jaws have evolved separately more than once. For instance, they occur in the southern salmonid fish (*Retropinna anisodon*) and in the iniomous fishes, which are probably a monophyletic group. Two main kinds of protractile jaws, atheriniform and acanthopterygian, were convergently evolved from maxillae-excluded jaws. In turn, the jaws of these two groups converge with those of cyprinoid fishes, whose ancestors presumably had jaws like those of primitive teleosts.

These convergences reflect the limitations inherent in the making of protractile mechanisms from a few parts. For analysis of these mechanisms in terms of machine kinematics, Alexanders' (1967) paper should be consulted. But to complete this survey, we should note that yet another kind of jaw protrusibility is found in the ribbon fishes and their allies (Allotriognathi). In most members of this order it is not the premaxillae that have processes to slide over the ethmoid part of the skull but the maxillae (Regan, 1907). Such jaws have not been "copied" by other fishes. If they are compared to the other four kinds of protractile jaws (phractolaemid, cyprinoid, atherinomorph, and acanthopterygian) it would seem that all possible protractile variations have been played on a primitive theme. The convergent variations—and they occur in the most successful groups—are simply more alike than the other two.

A most remarkable convergence, this time between structures involving a moderate number of parts, is so obvious that it might easily be overlooked. The limbs of land vertebrates were evolved from the pectoral and pelvic fins of rhipidistian species of tassell-finned (Crossopterygian) fishes. Why should there be five digits in both fore and hind limbs? One has to assume that pentadactyl limbs proved to be the most efficient for movement on land. There is also Smith's (1965) interesting suggestion. The cells of foams tend to have pentagonal faces, and it is this inherent, statistical feature of their topology that Smith believes may well account for the five-fold symmetry of plants and the pentadactyl limbs of animals.

CONVERGENCE AND LIVING SPACE

Though each kind of organism has its own living space, there are few main types of environment. On land, Elton (1966) classifies habitats into seven systems, which include organisms living between low-tide marks and the upper reaches of the atmosphere. Where suitable, these systems are divided laterally into formation types and vertically into certain layers. In the free space of the oceanic province there are four main habitat systems: epipelagic, mesopelagic, bathypelagic, and benthopelagic. Laterally, these may be divided broadly into polar, temperate, and tropical types, which divisions also apply to the benthic system of the deep sea. Vertically, the benthic system may be split into two main zones, slope (archibenthic) and abyssal (Ekman, 1953). The neritic province, divided vertically into intertidal and shelf systems, also has polar, temperate, and tropical subdivisions. In ponds and lakes, there is a littoral zone, a limnetic zone, and a profundal zone; and in streams two major parts may be recognized, a rapids zone and a pool zone (Odum, 1959).

Each habitat system poses its own constraints: physical, chemical, and biological. These are the Darwinian conditions of existence, which Cuvier saw in another light. To quote Russell (1946), Cuviers' idea of the conditions of existence was "that the organism must be a harmoniously working structuro-functional unity, capable of performing the metabolic and other functions which are essential to its life." But we have already given some thought to this aspect so far as it relates to convergence—under general properties of the medium and structural constraints. Here then, we will consider convergence in relation to Darwinian conditions of existence. Before we do so, one general property of environments and organisms is worth note. If each main kind of environment has its own adaptive types, which may be strongly convergent, the very fact that such types can be recognized is a reflection of the homeostatic properties of these environments (see also Pantin, 1965).

Since convergences are elicited by similar environmental conditions, may we then say that the extent of convergence is related to the severity level of these conditions? After observing that where selection pressure is high, animals tend to assume the same designs (Principle of Convergence), Rosen (1967) argues that there is a close relationship between selection pressure in an environment and the level of convergence in the inhabitants.

Of all the adaptive types considered earlier, the general level of convergence is highest in bathypelagic fishes. To recall previous discussion, these convergences involve their coloration, sense organs, sexual dimorphism, and their general organization. These similarities are related to such factors as the lack of sunlight, slow oceanic circulation, poor food supply, high predation pressure, adequate communication between the sexes, and the production of enough recruits to local populations.

There is regressive convergence between cave fishes, whose surroundings certainly resemble those of bathypelagic fishes in the lack of sunlight and sometimes in containing poor supplies of food. Cave fishes converge in developing regressed eyes and in strongly tending to lose their dermal pigmentation and scales.

Concerning the amblyopsid fishes, Poulson (1963) attributes their adaptations largely to a lack of predators and food scarcity, response to the latter being lowered metabolic and growth rates. Even if we consider only food-poor caves the level of convergence between cave fishes is not so pervasive as that reached by bathypelagic species. And while it is not easy to compare the levels of selection pressure in two environments, it would still be fair to say that such pressure in the bathypelagic zone is much higher than that in cave waters. So far then, there is some evidence to relate degrees of selection and convergence.

The Life of Fishes

Are selection pressures high in the habitats occupied by mormyrid and gymnotid fishes, which are strongly convergent? Most of these fishes are nocturnal, and they all produce series of weak electric fields, which they use for navigation, finding prey, predators, competitors, and so forth. The main selection pressures in their environments would thus seem to be related to such uses of their electrolocation systems. The surroundings of these fishes are surely more benign than those of bathypelagic fishes. But their many convergences, which involve the fine structure of the skin, the form of electroreceptors, the great development of the valvula cerebelli, the mode of swimming, body form, and so forth, simply follow from their evolution of electric organs. In other words, their convergences are largely and secondarily related to an environment of their own making.

Earlier, I touched briefly on another set of pervasive convergences: that linked to existence between grains of sand. Even the smallest fishes are much too large for the interstitial fauna. Indeed, the representatives of the many major groups that have found niches in sand have had to adjust their size, usually by regression, to a length of about a millimeter. They can thus move between the grains. Hence, also, the strongly contractile body wall and the elongated shape of many interstitial animals. It would seem, then, that the main selection pressures are associated with the small labyrinthine complex of watery spaces between the grains of sand. Supplies of food may be quite favorable judging by observations on New Zealand shores. Morton and Miller (1968) write: "The primary industry of photosynthesis is carried out in the interstitial community by diatoms and peridinian dinoflagellates which may migrate upwards in immense numbers as an evanescent oily surface layer. Many herbivores browse on these plants, such as ostracods, copepods and small worms and the nudibranch molluscs . . . Others such as nematodes, gastrotrichs, and tardigrades suck out their contents with minute pumping devices. Cumacean and amphipod crustaceans lick the sand grains for bacterial and other surface growth. A number of worm species are general detritus feeders and in the hydroid polypes and turbellarians we have carnivores represented."

Consideration of the interstitial fauna might lead to this surmise: the diversity of animals involved in convergences is highest in environments where selection pressures are very strong. We might be reminded of life in torrents, where the force of the current and the finding of enough food are searching selection pressures. Many forms, including fishes, browse on the algae and slime that cover the rocks and stones. Some forms, such as flatworms and snails cling to rocks by means of their sticky undersurfaces. Leeches, which have suckers, are to this extent preadapted for life in torrents, but large suckers of one sort or another have evolved in tadpoles, fishes, and various insects. Most torrential animals, from insect larvae to fishes have flattened bodies with a streamlined profile. Many are positively rheotactic and thigmotactic. If we compare the faunas of the rapids and slow-flowing zones of streams, it would be fair to say that more distantly related groups of animals are convergent in the former zones.

But how many groups of bathypelagic animals show fish-like convergences? The cephalopods, which have reduced their tissue systems and developed much gelatinous padding, have "copied" the fishes. So have some of the decapod crustaceans, such as *Notostomus* spp. Like the fishes, these two groups had to evolve low-cost tissue systems to fit a food-poor environment. But the small crustaceans do not seem to be reduced in this way. Even the bathypelagic euphausiids, such as *Bentheuphausia* and *Thysanopoda* spp., show no marked tissue economies, except for their reduced eyes, when compared to mesopelagic and epipelagic species. Perhaps the smaller crustaceans, because of their system of limbs, bristles, and food-collecting devices are better able than fishes or cephalopods to make a living in food-poor surroundings. Yet many of the bathypelagic copepods are carnivores rather than filter feeders. At all events, here is an environment where high selection pressures are related mainly to a scarcity of food and where corrlated convergences involve only the largest and most complex forms. Members of the interstitial fauna, though very diverse, are necessarily all much the same size. Thus, our earlier surmise has its limitations.

INTERNAL CONSTRAINTS

Do some convergences arise because only a few genetic and developmental processes are viable? Whyte (1965a) insists on the prevalence of internal selection, "the restriction of the direction of evolutionary change by internal organizational factors, i.e., selective processes acting directly on the early consequences of the genotype which ensure that the co-ordinative conditions are satisfied by all mutated types that survive up to the point at which Darwinian external selection enters." Bernal (1967) refers to this hypothesis when discussing his ideas of the inner, prescribed molecular nature of life. But the opening question can only be posed at the present time, whereas the genetic background of divergence is more easily studied.

THE RECOGNITION OF CONVERGENCE AND PHYLETIC STUDIES

Scientific theories lend themselves to prediction, but this is not true of evolutionary theory. Divergences and the consequences of such, certainly cannot be foretold. But assuming natural selection to shape life into nothing but a complex of adaptations, a biologist with an intimate knowledge of selective processes and great powers of deduction, could, according to MacArthur and Connell (1966) "predict quite closely the form and function of each phenotypic trait." After listing four reasons why prediction must be uncertain (lack of time for development of a character, random mortality, limited kinds of mutations, and the necessity of building on the present gene complex), these two authors continue: "The best evidence that prediction is possible comes from the marvellously close convergence which is frequent between unrelated organisms adapted to similar ways of life." In practice, then, convergences may be predictable; divergences are not. For instance, if we found two distantly related fishes in the early stages of assuming life in cave waters, we could predict that their eyes, pigment, and scales would eventually regress. But, prediction would be less sure for two such fishes assuming life in swamps. Presumably both would be preadapted for life in stagnant waters on account of their facility for nibbling the oxygenated water close to the surface. Life in swamps might simply lead to an improvement of this habit, not necessarily to the evolution of an air-breathing organ.

Evolution is a stochastic process, which means it is not repeatable. Dollos' law of evolutionary irreversibility, proposed by Abel in 1912, was the first, though not altogether precise, intimation of this fact. But it does mean that once a structure is lost it is never regained in identical form. The evolution of a structure represents a long chain of events and circumstances, which are most unlikely to be repeated. Hence lineages and their characters are unique. We appreciate this when faced with convergence. Close study of convergent designs reveals that they differ in structural details, which are marks of their uniqueness.

But at microscopic levels, convergent structures may display elaborate similarities. Earlier, we mentioned that ciliary patterns in certain filter-feeding invertebrates almost look like homologies. A further pseudohomology concerns the excretory organs of cephalochordates and annelid worms. Goodrich (1902) was once so impressed with the similarities between these organs that he was led to write: "If two such excretory organs as the solenocyte-bearing nephridia of *Phyllodoce* and the solenocyte-bearing kidneys of *Amphioxus* could be shown to have been involved, we should have to give up structural resemblance as a guide to homology." The two genera in question were surely evolved independently, and the close similarity between their excretory organs is confined to their solenocytes, found also in Gastrotricha and Kinorhyncha. Solenocytes are thought to be a variant form of flame-bulbs, which occur in flatworms, rotifers, and nemertean worms. Evidently, the action of the bundles of cilia ("flames") or a single cilium is to drive excretory products down the tubule. How flame-bulbs or solenocytes form urine is unknown. When we do know their function, we may realize how "inevitable" is the convergent evolution of solenocytes and flame-bulbs.[18] If so, we may also be reminded of the essential similarity in fine structure of tubular cells that create differences in osmotic pressure in excretory organs. In mammals, snails, and crayfish, for instance, these cells have complex infoldings of the basal membrane with many associated mitochondria (Schmidt-Nielsen, 1965).

If the aim of classification is to reveal the evolutionary relationships of organisms, care must be taken to avoid convergent characters. Thus, if the extent of convergence

18. There is also close convergence between the fine structure of the eyecups of Hesse in *Amphioxus* and the prostostomial ocelli of the opheliid worm, *Armandia brevis* (Hermans and Cloney, 1966).

is not to be underestimated, or overestimated, the classifier should at least take trouble to cover the functional morphology and ecology of the groups under consideration. Adaptive characters may be suspect. More precisely, the most useful characters are those that are unlikely to be affected by selection pressure. Contrarily, characters that closely "represent the environment," to use Young's (1964) phrase, are particularly prone to convergence. For instance, we have seen how detailed and pervasive are the convergences between the sense organs of fishes. Characters that camouflage fishes may also be subject to elaborate convergence.

If camouflage tends to make a fish vanish from view in a particular environment, the swimbladder also "represents the environment," this time in bringing a teleosts' weight in water to the vanishing point. Cope classified the teleosts into physostomes and physoclists, but we saw that a closed swimbladder has evolved convergently several times. On the other hand, once a teleost has acquired certain requisite structures for maintaining neutral buoyancy, the pattern of these structures should be independent of selection pressure and thus usable in phyletic classification. For instance, the bauplan of resorbent and gas-producing tissues is so alike in albuloids, halosaurs, notacanths, and eels that one is led to think of a common ancestor for these groups (see also Marshall, 1962).

Recently, Greenwood, *et al.* (1966) combined the whalefishes (Cetunculi) and Miripinnati in one order, Cetomimiformes. Support for this decision comes from the similar swimbladder bauplan of a whalefish (*Barbourisia* and that of the Miripinnati. In both, the gas gland, which is in the forepart of the sac, is fed through two unipolar retia mirabilia that run forward from an origin in the middle region of the sac. The resorbent part of the swimbladder is immediately behind the gas-producing section.

Body form, locomotor organs, and fin patterns, particularly as they are correlated with predation, escape, or maneuver, are certainly subject to strong selection pressure. Thus, we saw that small surface-feed characins look very like cyprinodont fishes, which are also surface feeders. If these distantly related forms show much convergence, how true may this be of the synentognath fishes, atherinids, and cyprinodonts, which again tend to feed at the surface. Rosen (1964) puts these three groups in one order, Atheriniformes; but Gosline (1968) believes that the resemblances between them are due to convergence.

Last, structures involved in feeding are likely to be strongly influenced by selection pressure and thus be prone to convergence. Convergence in jaw protrusibility have already been considered, and there are even more striking convergences in tooth form and pattern. On the other hand, epibranchial organs, which are found in microphagous fishes, are so closely alike that they were probably inherited from a single ancestor.

CONVERGENCES BETWEEN NATURAL AND MAN-MADE OBJECTS

From time to time men discover that they have designed objects with certain similarities to natural structures. One such discovery is told by D'Arcy Thompson (1961, p. 232): "A great engineer, Professor Culman of Zurich, to whom, by the way, we owe the whole modern method of 'graphic statics,' happened (in the year 1866) to come into his colleague Meyer's dissecting-room where the anatomist was contemplating the section of a bone. The engineer, who had been busy designing a new and powerful crane, saw in a moment that the arrangement of the bony trabeculae was nothing more nor less than a diagram of the lines of stress, or directions of tension and compression, in the loaded structure: in short, that Nature was strengthening the bone in precisely the manner and direction in which strength was required; and he is said to have cried out, 'That's my crane!'" (see also Murray, 1936). More recently, Buckminster Fuller (1965), after designing geodesic structures for many years, was delighted to find them in nature: in the protein shells of viruses, in diatoms, radiolarians, and so forth. Again, after designing target-seeking equipment, Wiener was led to look for analogous system in nature, that is, for those involving negative feedback.

Now there is a science of bionics, which involves the design of physical systems by analogy with living systems. Thus, rather than "discover" the independent development of natural devices and his own, man is now looking to nature for inspiration and suggestion (Gérardin, 1968).

WIDER ASPECTS

Biologists are not the only scientists concerned with convergent phenomena. Langer (1967) refers to Heisenberg's observation that impressions of colors may be related to the atomic structure of chemicals, to the refraction of light, or to retinal or cerebral processes. She quotes Heisenberg's conclusion: "We see here a general characteristic of nature. Processes appearing to our senses to be closely related often lose this relation when their causes are investigated." Yet it is in the social sciences that we come closest to the biologist's sense of convergence. For instance, when social anthropologists study the creations of native peoples, they may be led to consider whether the similarities between the products of two separate peoples is related to diffusion from a common source or to some form of convergence (called parallelism by some authorities). There are certain constraints to be kept in mind. Are we sure, for instance, that knowledge of wheel making spread from one center only? Might not the idea have occurred to more than one people, provided that they lived in a region where, for instance, log rolling was possible? Again, are we sure that percussion instruments were invented once and only once? What about woodwind instruments? Concerning weapons, did only one set of people think of making a spear, or even bows and arrows? Where there is restriction on the ways materials can be used for certain ends, there is always a strong chance of convergent creation. This is also true, as we saw of biological convergence.

Of man's systems of symbols, Meerloo (1966) observes: "Just as the history of evolution is condensed and stored in the genes as bearers of genetic memory, the history of human thinking and communication is condensed and stored in human symbols." That most far-reaching system of signs, the alphabet, evidently came from one source (Diringer, 1968). One reason is that the script used by Semitic-speaking Syrians and Palestinians (second half of the second millennium B.C.) shows close resemblances, even detailed ones, to much later Phoenician and early Hebrew inscriptions (first half of the first millennium B.C.). In an analogous way, evolutionary lines may be traced through the persistence of idiosyncratic features. On the other hand, certain nonalphabetic systems of signs which seem to have been invented independently, contain similar inscriptions. Is it surprising, though, that two separate peoples should think independently of using a wavy line as a symbol for water?

At a higher symbolic level, recent discussion between Chomsky and Hampshire (1968) on linguistic studies is apt. Chomsky stresses two aspects of grammars: (1) their creative nature, whereby they generate an indefinitely large number of sentences; and (2) their restrictive side (evidence suggests that these are innate factors, probably of a highly restrictive nature, that determine how knowledge of any language emerges in the individual, given the limited data available to him). Again, there are constraints and they must be remembered in comparative studies. For instance, "Very extensive studies would be required to show how far the recurrence of metaphors is due to diffusion from a single ancient source and how far due to a natural tendency in man to invent the same metaphors independently" (Madge, 1963).

Social anthropologists—and their structural studies owe much to structural linguistics—often have to balance diffusion against convergence. Lévi-Strauss (1963) gives a stimulating critique of split representation in the art of Asia and America. For instance, there are many similarities between such art in China (first to second millennia B.C.) and that of Alaskan Indians (eighteenth and nineteenth centuries A.D), similarities so close that the technological and artistic principles shown by both are almost entirely identical. This resemblance might be due to diffusion. But are the many similarities between Alaskan and Maori art to be explained in the same way? As Lévi-Strauss says: cultural contact is the most satisfactory hypothesis to account for complex similarities that chance cannot explain. If contact is eventually ruled out by historians, there must be some other explanation. Constraints must be considered, such as the limitations imposed in making split designs on masks, human faces, boxes, bracelets, and so forth. This is not the place to follow all of Lévi-Strauss' arguments, but he implies that there is not enough evidence to resolve the problem of diffusion versus "convergence." For our purpose, though, we may simply note the parallels between the problems of social anthropologists and biologists.

In the widest sense, then, convergence must be considered by students of human culture as well as by biologists. Biologists are now concerned with structure rather than form; structure as Whyte (1965b) says, being "form seen inside." Students of human culture now tend to think in structural terms. Indeed, it is now possible to think generally of "l'activité structuraliste" (see Barthes, 1964). Is it surprising, then, that we see more clearly the common ground between students of biological and psychosocial evolution?

Appendix   Classification and Evolution of Teleost Fishes

In Figure 1, p. 2, is shown the evolutionary relationships of the main groups of fishes. This brief outline of teleost radiation is intended to help non-specialists, particularly in reading Chapters 1 and 5.

More is owed to C. Tate Regan than to any other single ichthyologist in bringing order to the complex of teleost fishes. Regan outlined his classification under "Fishes" in the fourteenth edition (1929) of the Encyclopaedia Britannica. Many of the most primitive teleosts—those most like their holostean ancestors—are placed in the order Isospondyli with suborders Clupeoidea (including elopids, albulids, alepocephalids, and *Chanos* as well as the herring-like fishes), Salmonoidea, Stomiatoidea, Osteoglossoidea, Mormyroidea, Notopteroidea, and Gonorhynchoidea. The other order containing primitive forms, particularly among the characids, is the diverse and predominantly freshwater group Ostariophysi (characids and carp-like fishes (Cyprinoidea) and catfishes (Siluroidea). The unique feature of the Ostariophysi is the chain of ossicles linking the swimbladder to the inner ears (Weberian apparatus). The most primitive members of these two orders "can be recognized, *inter alia*, by their fin pattern, jaw structure, scaling and swimbladder. There is a single dorsal fin, a symmetrical tail fin typically formed of nineteen principal rays, and a single anal fin. The pectoral fins are set low down on the shoulders and the pectoral girdle is braced on each side by an inner strut of bone called the mesocoracoid. The pelvic fins, which have numerous rays, are inserted on the abdomen, well back from the pectorals. There are no fin spines, the fin rays being jointed and soft. The biting part of the upper jaw is formed by a small premaxilla and a long blade-like maxilla. The scales are cycloid, that is without teeth on the free edges. Lastly, the swimbladder opens by way of a pneumatic duct into the roof of the foregut" (Marshall, 1965).

Regan's other main orders are:

Iniomi (Aulopidae, Synodontidae [lizard-fishes], lantern-fishes, alepisauroids, ateleopids, giganturids, and so on). They differ from isospondylous fishes in developing premaxillae that exclude the maxillae from the gape, and so on.

Apodes (eels). They are elongated fishes with soft rays, long dorsal and anal fins, small caudal fin, no pelvic fins, small gill openings, a leptocephalus larval phase, and so on.

Heteromi (halosaurs and notacanths). They are bottom-dwelling, deep-sea fishes with a closed swimbladder, a long, tapering tail with a long anal fin and no caudal fin.

Synentognathi (needle-fishes, sauries, half-beaks and flying-fishes). Features are closed swimbladder; abdominal pelvic fins, lower pharyngeal bones united, dorsal fin far back above the anal fin; lateral line set near lower edge of body, and so on.

Microcyprini (blind cave-fishes [amblyopsids] and cyprinodonts). They are soft-rayed fishes with abdominal pelvic fins, a closed swimbladder, separate lower pharyngeal bones, and so on.

Anacanthini (hakes, cod-fishes, rat-tails, and so on). They are soft-rayed fishes with numerous fin rays, pelvic fins set below or in advance of the pectorals, caudal fin reduced or absent, swimbladder closed, usually with an "oval" resorbent area, and so on.

Solenichthyes (tube mouths, *Aulostomus*, *Fistularia*, snipe-fishes, pipe-fishes, sea-horses, and so on). The mouth is at the end of a long tube-like snout; pelvic fins are abdominal when present; swimbladder is closed, usually double-chambered, and so on.

Berycomorphi (polymixiids, berycids, melamphaids, holocentrids [squirrel-fishes], and so on). Scales are usually toothed; anterior rays of dorsal and anal fins are spiny; caudal fin usually has 19 principal rays; mouth is bordered by protractile premaxillae, and so on.

Percomorphi (perch-like fishes [Percoidea], acanthuroids [surgeon-fishes] scombroids [mackerel and tuna], gobioids, blennioids, anabantoids, stromateoids, and so on. "This large order, which contains several thousand species, is difficult to define. It includes the typical Perches, which originated from the Berycoids at the end of the Cretaceous period. They are fishes of normal form, with a spinous dorsal fin, thoracic pelvic fins, each of a spine and five branched rays, pelvic bones attached to the pectoral arch, a caudal fin with seventeen principal rays, and the mouth bordered above by the pro-tractile premaxillaries. But other fishes of the order may be of various shapes, may have no spinous fin-rays, may have the pelvic fins farther back and their bones uncon-nected with the pectoral arch, or placed on the throat, or reduced or absent; some have no caudal fin, some a non-protractile mouth, etc., etc. In other words, many diverse types are included, either because they resemble the typical perches in many of their characters, or are con-nected with them by transition-forms" (Regan, 1936).

Scleroparei (mail-cheeked fishes; scorpion-fishes [Scorpaenidae], bull-heads [Cottidae], gurnards [Trig-lidae], sea snails [Liparidae], and so on). They are closely related to percomorphs but have the second suborbital bone produced as a strut across the cheek to support the preopercular bone.

Heterosomata (flat-fishes; flounders, soles, and so on). They have both eyes on the same side of the head, the body strongly compressed with long, many-rayed dorsal and anal fins; the swimbladder is absent, and so on.

Plectognathi (trigger-fishes, file-fishes, puffer fishes, porcupine fishes, ocean sunfishes, and so on). They re-semble percomorphs but are distinguished by the short and powerful jaws bearing a few strong teeth, small gill openings, bony or spiny scales, and so on.

Xenopterygii (cling fishes). The pelvic fins form a large suction-cup disc on the flattened underparts of the trunk; dorsal and anal fins have few rays, and so on.

Pediculati (angler-fishes). Spinous dorsal ray of a few flexible rays, the first set on top of the head and used as a lure, pectoral fins with a basal lobe, gill openings small, and so on. In evolutionary terms this classification has been seen thus: evolutionary radiations of a primitive isospondylous stock not only gave rise to the Isospon-dyli but to the orders Ostariophysi to Berycomorphi. The Berycomorphi and the percomorphs had a common ori-gin. From primitive percomorph ancestors radiated the orders Scleroparei to Pediculati.

Recent phyletic studies (Rosen, 1964, Greenwood *et al.*, 1966, and Rosen and Patterson, 1969, have done much to improve conceptions of teleost evolution. Parts of the order Isospondyli, which appeared to some as a polyphyletic assemblage, are seen to form evolutionary trends represented by the superorders Elopomorpha (eels, notacanths, halosaurs, and gulper-eels as well as the elopids and albulids), Clupeomorpha, and Osteoglossomorpha (osteoglossoids, notopteroids, and mormyroids). The salmonoid, stomiatoid, gonorhychoid and chanoid groups of Isospondyli, together with such orders as the Haplomi and Iniomi are placed in the superorder Protacanthopterygii. The Ostariophysi are elevated to superordinal rank. Another trend in teleost evolution is expressed in the superorder Paracanthopterygii, which includes the amblyopsids trout-perches, toad-fishes, cling-fishes, anglerfishes and the Anacanthini (expanded to contain the ophidioid and zoarcoid fishes). The cyprinodonts, atherinids, and Synentognathi are classified in the superorder Atherinomorpha. Finally, the Berycomorphi, Percomorphi, Scleroparei, Heterosomata, and Plectognathi form most of the great spiny-finned superorder Acanthopterygii.

The reasons behind these recent conceptions are too lengthy to be considered here, and the reader must refer to the original publications. In evolutionary terms the pholidophoroid holosteans are considered to have given rise separately to elopomorph, osteoglossomorph, protacanthopterygian, and possibly clupeomorph trends. The paracanthopterygian and acanthopterygian stocks are thought to come from the Protacanthopterygii, the first via some neoscopelid-like fish, the second from the ctenothrissiforms. The atherinomorphs and Ostariophysi may also have been derived from protacanthopterygian ancestors (Greenwood et al., 1966).

These studies imply that the teleosts are a polyphyletic assemblage, though Patterson (1967) concludes that the transition from pholidophoroids to teleosts was across a "narrow front."

References

Alexander, R. McN. (1965) "Structure and function in the catfish." *J. Zool. Lond.* **148**:88–152.

—— (1966) "Physical aspects of swimbladder function." *Biol. Rev.* **41**:141–176.

—— (1967a) "Mechanisms of the jaw of some atheriniform fish." *J. Zool. Lond.* **151**:233–255.

—— (1967b) *Functional design in fishes.* Hutchinson, London.

Andriashev, A. P. (1964) "Fishes of the northern seas of the U.S.S.R." *8° Israel Prog. Sci. Trans. Jerusalem.* 1–617.

Angelis, C. de (1950) "Le cellule di Mauthner in alcuni Teleostei di fondo." *Boll. Pesc. Piscic. Idrob.* **26**:256–263.

Aronson, L. R. (1963) "The central nervous system of sharks and bony fishes with special reference to sensory and integrative mechanisms." In *Sharks and survival,* ed. Perry W. Gilbert, pp. 165–241. D. C. Heath and Company, Boston.

Von Arx, W. S. (1962) *An introduction to physical oceanography.* Addison Wesley Publishing Co.

Backus, R. H. (1968) "Solving the mystery of Alexander's Acres." *Oceanus* **14** (3):15–20.

Bainbridge, R. (1960) "Speed and stamina in three fish." *J. exp. Biol.* **37**:129–153.

—— (1963) "Caudal fin and body movement in the propulsion of some fish." *J. exp. Biol.* **40**:23–56.

Baker, A. de C. (1963). "The problem of keeping planktonic animals alive in the laboratory." *J. mar. biol. Ass. U.K.* **43**:291–294.

Bănărescu, P. (1957) "Vergleichende Anatomie und Bedeutung der *Valvula cerebelli* der Knochenfische." *Rev. Biol. Bucureşti* **2**:255–276.

Bardack, D., and R. Zangerl (1968) "First fossil lamprey: A record from the Pennsylvanian of Illinois." *Science* **162**:1265–1267.

Bardach, J. E., J. H. Todd, and R. Crickmer (1967) "Orientation by taste in fish of the genus *Ictalurus*." *Science* **155**:1276–1268.

Barets, André (1961) "Contribution à l'étude des systèmes moteurs 'lent' et 'rapide' du muscle latéral des teleostéens." *Archs. Anat. microsc. Morph. exp.* **50** (1):92–187.

Barham E. G. (1966) "Deep scattering layer migration and composition: Observations from a diving saucer." *Science* **151** (3716):1399–1403.

Barthes, Roland (1964) *Essais critiques,* Paris, Éditions du Seuil.

Bates, Marston (1960) *The forest and the sea.* New York, Mentor Book, 1961.

The Life of Fishes

Békésy, G. von (1967) "Some similarities in sensory perception of fish and man." In *Lateral line detectors*, ed. Phyllis H. Cahn, pp. 417–435. Bloomington, Indiana University Press.

Belyaev, G. N. (1966) "Bottom fauna of the ultraabyssal of the world ocean." Second International Oceanographic Congress, Nauka, Moscow, pp. 32–33. Abstract.

Berg, T., and J. B. Steen (1966) "The gills of two species of haemoglobin-free fishes compared to those of other teleosts—with a note on severe anaemia in an eel." *Comp. Biochem. Physiol.* **18** (3):517–526.

Bernal, J. D. (1967) *The origin of life*. London, Weidenfeld and Nicolson.

Bertelsen, E. (1951) "The ceratioid fishes." *Dana Rep.* no. 39:1–276.

Bertelsen, E., and O. Munk (1964) "Rectal light organs in the argentinoid fishes *Opisthoproctus* and *Winteria.*" *Dana Rep.* no. 62:1–17.

Bianki, V. C. (1963) "Swimbladder receptor function and the cerebellum." *Fiziol. Zh. S.S.S.R.* **49**, 494: (1964) *Fedn. Proc. Fedn. Am. Socs. exp. Biol.* **23**, T. 222–T. 226.

Bierbaum, G. (1914) "Untersuchungen über den Bau der Gehörorgane von Tiefseefischen." *Z. wiss. Zool.* **III**:281–380.

Blaxter, J. H. S., and R. I. Currie (1967) "The effect of artificial lights on acoustic scattering layers in the ocean." *Symp. zool. Soc. Lond.* no. 19:1–14.

Boddeke, R., E. J. Slijper, and A. van der Stelt (1959). "Histological characteristics of the body-musculature of fishes in connection with their mode of life." *Proc. Ned. Akad. Wet. Amst.* **62C**:576–588.

Bone, Q. (1966) "On the function of two types of myotomal muscle fibre in elasmobranch fish." *J. mar. Biol. Ass. U.K.* **46**:321–349.

Bonner, J. T. (1965) *Size and cycle: An essay on the structure of biology*. Princeton University Press.

Brittan, M. R. (1961) "Adaptive radiation in Asiatic cyprinid fishes, and their comparison with forms from other areas." *Proc. 9th Pacific Science Congress*, **10** (Fisheries):18–31.

Brauer, A. (1908) "Die Tiefsee-Fische" *II Anatomische Teil. Wiss. Ergebn. "Valdivia"* **15** (2):1–266.

Breder, C. M. (1926) "The locomotion of fishes." *Zoologica* **4**:159–297.

———(1932) "On the habits and development of certain Atlantic Synentognathi." *Publs. Carnegie Instn.* **435**:1–35.

——— (1934) "The reproductive habits of the painted *Betta,* a relative of the Siamese fighting fish, new to aquaria." *Bull. N.Y. zool. Soc.* **37**:126–133.

——— and D. E. Rosen (1966) *Modes of reproduction in fishes*. New York, Natural History Press.

Bruun, A. Fr. (1957) "Deep sea and abyssal depths" Geol. Soc. America. Memoir 67. Vol. 1:641–672.

Burne, R. H. (1909) "The anatomy of the olfactory organ of teleostean fishes." *Proc. zool. Soc. Lond.*:610–663.

References

Butler, C. G. (1967) "Insect pheromones." *Biol. Rev.* **42** (1): 42–87.

Carey, F. G., and J. M. Teal (1966a) "Heat conservation in tuna fish muscle." *Proc. natn. Acad. Sci. U.S.A.* **56**: 1464–1469.

———(1966b) "Why is a tuna warm?" *Oceanus* **13** (1): 8–11.

———(1969) "Regulation of body temperature by bluefin tuna." *Comp. Biochem. Physiol.* **28**: 205–213.

Cattell, M. (1936) "The physiological effects of pressure." *Biol. Rev.* **2**: 441–476.

Chomsky, Noam, and S. Hampshire (1968). "Discussion on the study of language." *Listener B.B.C.* **79** (2044): 687–691.

Clarke, G. L., and R. H. Backus (1964) "Interrelations between the vertical migration of deep scattering layers, bioluminescence and changes of daylight in the sea." *Bull. Inst. oceanogr. Monaco.* **64** (1318): 1–36.

Clarke, G. L., and M. G. Kelly (1964) "Variation in transparency and in bioluminescence on longitudinal transects in the Western Indian Ocean." *Bull. Inst. oceanogr. Monaco* **64** (1319): 1–20.

Clarke, W. D. (1963) "Function of bioluminescence in mesopelagic organisms." *Nature* 198 (4887): 1244–1246.

Denton, E. J. (1960) "The 'design' of fish and cephalopod eyes in relation to their environment." *Symp. zool. Soc. Lond.* no. 3: 53–55.

(1963) "Buoyancy mechanisms of sea creatures." *Endeavour* **22**: 3–8.

Denton, E. J., and F. J. Warren (1957) "The photosensitive pigments in the retinae of deep-sea fish." *J. mar. biol. Ass. U.K.* **36**: 651–662.

Denton, E. J., and N. B. Marshall (1958) "The buoyancy of bathypelagic fishes without a gas-filled swimbladder." *J. mar. biol. Ass. U.K.* **36**: 753–767.

Denton, E. J., and J. A. C. Nicol (1965) "Studies on reflexion of light from silvery surfaces of fishes, with special reference to the bleak, *Alburnus alburnus.*" *J. mar. Biol. Ass. U.K.* **45**: 683–703.

———(1966) "A survey of reflectivity in silvery teleosts." *J. mar. Biol. Asso. U.K.* **46**: 685–722.

Denton, E. J., J. B. Gilpin-Brown, and J. V. Howarth (1967) "On the buoyancy of *Spirula spirula.*" *J. mar. biol. Ass. U.K.* **47**: 181–191.

Digby, P. S. B. (1967) "Pressure sensitivity and its mechanism in the shallow marine environment." *Symp. zool. Soc. Lond.* no. 19: 159–188.

Dijkgraaf, S. (1960) "Hearing in bony fishes." *Proc. Roy. Soc. B.* **152**: 51–54.

——— (1962) "The functioning and significance of the lateral-line organs." *Biol. Rev.* **38**: 51–105.

Diringer, David (1968) *The alphabet.* Vols. 1 and 2. London, Hutchinson.

Ebeling, A. W. (1962) "Melamphaidae 1: Systematics and zoogeography of the species in the bathypelagic fish genus *Melamphaes* Günther." *Dana Rep.* no. 58: 164.

Ekman, S. (1953) *Zoogeography of the sea.* London, Sidgwick and Jackson.

Elton, C. S. (1966) *The pattern of animal communities.* London, Methuen.

Engström, K. (1963) "Cone types and cone arrangements in teleost retinae." *Acta Zoologica* **44** (1–2):179–243.

Fahlén G. (1965) "Histology of the posterior chamber of the swimbladder of *Argentina.*" *Nature,* Lond. **207**:94–95.

—————— (1967) "Morphological aspects on the hydrostatic function of the gas bladder of *Clupea harengus L.*" *Acta. Univ. Lund.*:1–49.

Fänge, R. (1953) "The mechanism of gas transport in the euphysoclist swimbladder." *Acta physiol. Scand.* **30** *Suppl. 110*: 1–133.

Fierstine, H. L., and V. Walters (1968) "Studies in locomotion and anatomy of scombroid fishes." *Mem. S. C. Acad. Sci.* **6**:1–31.

Filatova, Z. A., and L. A. Zenkevitch (1966) "The quantitative distribution of deep-sea bottom fauna in the Pacific Ocean." Second International Oceanographic Congress, Nauka, Moscow, pp. 114–115 Abstract.

Flock, Å. (1965) "Electron microscopic and electrophysiological studies on the lateral line canal organ." *Acta Oto-laryngologica suppl. 199*:1–90.

Ford, E. (1937) "Vertebral variation in teleostean fishes." *J. mar. biol. Ass. U.K.* **22**:1–60.

Foxton, P. (1964) "Observations on the early development and hatching of the eggs of *Acanthephyra purpurea* A. Milne-Edwards." *Crustaceana* **6**:235–237.

Fraser, J. H. (1962) *Nature adrift: The story of marine plankton,* London, G. T. Foulis.

Freihofer, W. C. (1963) "Patterns of the Ramus Lateralis Accessorius and their systematic significance in teleostean fishes." *Stanford Ichthyol. Bull.* **8**:81–189.

Fridberg, G., and H. A. Bern (1968) "The urophysis and the caudal neurosecretory system of fishes." *Biol. Rev.* **43** (2): 175–199.

Fuller, Buckminster R. (1965) "Conceptuality of fundamental structures." In *Structure in art and in science,* ed. Gyorgy Kepes, pp. 66–88. London, Studio Vista.

Gardner, Martin (1967) *The ambidextrous universe.* London, The Penguin Press.

Garman, S. (1899) "Reports on an exploration off the west coasts of Mexico, Central and South America and off the Galapagos Islands. The fishes." *Mem. Mus. comp. Zool. Harv.* **24**:1–431.

Gérardin, L. (1968) *Bionics.* London, Weidenfeld and Nicolson.

Gilbert, P. W. (1963). "The visual apparatus of sharks." In *Sharks and Survival,* ed. Perry W. Gilbert, pp. 283–326. Boston, D. C. Heath and Company.

Gilbert, C. H. and C. L. Hubbs (1920). The macrourid fishes of the Philippine Islands and the East Indies. *Bull. U.S. natn. Mus.* **100**, *vol* 1., *pt.* 7.

Goode, G. B. and T. H. Bean (1896) "Oceanic ichthyology" *Spec. Bull. U.S. natn. Mus.* no. 2, pp. 553.

References

Goodrich, E. S. (1902) "On the structure of the excretory organs of *Amphioxus*." *Quart. J. micros. Sci.* **45**:493–501.

Gosline, W. A. (1968) "The suborders of perciform fishes." *Proc. U.S. natn. Mus.* **124** (3647):1–77.

Gray, I. E. (1954) "Comparative study of the gill area of marine fishes." *Biol. Bull. Woods Hole,* **107**:219–225.

Gray, J. (1933) "Directional control of fish movement." *Proc. Roy. Soc. Lond. B.* **113**:115–125.

———(1953) "The locomotion of fishes." In *Essays in marine biology*, pp. 1–16. Oliver and Boyd, Edinburgh.

——— (1968) *Animal locomotion*. London, Weidenfeld and Nicolson.

Greenway, P. (1965) "Body form and behavioural types in fish." *Experientia* **21**:489–498.

Greenwood, P. H., D. E. Rosen, S. H. Weitzman, and G. S. Myers (1966) "Phyletic studies of teleostean fishes with a provisional classification of living forms." *Bull. Amer. Mus. Nat. Hist.* **131** art. 4:339–456.

Grenholm, A. (1923) "Studien über die Flossenmuskulatur der Teleostier." Inaug. Diss. Uppsala Univ. Arsskrift 1923 Matem. Naturvet. 2 Uppsala IX.

Haldane, J. B. S. (1928) *Possible worlds*. London, Chatto and Windus.

Harris, J. E. (1953) "Fin patterns and mode of life in fishes." In *Essays in marine biology*, pp. 17–28. Edinburgh, Oliver and Boyd.

Harrisson, C. M. H. (1967) "On methods for sampling meso-pelagic fishes." In *Aspects of marine zoology*, ed. N. B. Marshall. *Symp. Zool. Soc. Lond.* no. 19:71–126.

Haslewood, G. A. D. (1964) "The biological significance of chemical differences in bile salts." *Biol. Rev.* **39** (4):537–574.

Hermans, C. O., and R. A. Cloney (1966) "Fine structure of the protostomial eyes of *Armadia brevis* (Polychaeta:Ophelii-dae)." *Z. Zellforsch.* **72**:583–596.

Herrick, C. J. (1961) *The evolution of human nature*. New York, Harper.

Herring, P. J. (1967) "The pigments of plankton at the sea surface." *Symp. zool. Soc. Lond.* no. 19:215–236.

Hertel, H. (1966) *Structure, form, movement*. New York, Reinhold.

Hessler, R. R., and H. L. Sanders (1967) "Faunal diversity in the deep sea." *Deep-sea Res.* **14** (1):65–78.

Hiatt, R. W., and D. W. Strasburg (1960) "Ecological relationships of the fish fauna on coral reefs of the Marshall Islands." *Ecological Monographs* **30**:65–127.

Hjort, J. (1912) "General biology." In *The depths of the ocean*, ed. J. Murray and J. Hjort, pp. 660–785. London, Macmillan.

Hoar, W. S. (1966) *General and comparative physiology*. Englewood Cliffs, N.J., Prentice-Hall, Inc.

Hollister, C. D., and B. C. Heezen (1966) "Ocean bottom currents." In *Encyclopaedia of Oceanography*, ed. R. W. Fairbridge, pp. 576–583. New York, Reinhold.

177

Horridge, G. A. (1966) "Some recently discovered underwater vibration receptors in invertebrates." In *Some contemporary studies in marine science,* ed. Harold Barnes, pp. 395–405.

Horridge, G. A., and P. S. Boulton (1967) "Prey detection by Chaetognatha via a vibration sense." *Proc. Roy. Soc. B.* **168** (1013):413–419.

Hughes, G. M. (1960) "A comparative study of gill ventilation in marine teleosts. *J. exp. Biol.* **37**:28–45.

Hutchinson, G. E. (1958) "Homage to Santa Rosalia or Why are there so many kinds of animals." *Amer. Nat.* **93** (870): 145–159.

————(1965) "The influence of the environment." In *The scientific endeavour,* pp. 235–240. New York, The Rockefeller Institute Press.

————(1966) *A treatise on limnology.* Vol. 2 *Introduction to lake biology and the limnoplankton.* New York, John Wiley.

Hutchinson, G. E., and R. B. MacArthur (1959) "A theoretical ecological model of size distributions among species of animals." *Amer. Nat.* **93**:117–126.

Ishimaya, R. (1955) "Studies on the rays and skates belonging to the family Rajidae, found in Japan and adjacent regions. 5. Electric organ supposed as an armature." *Bull. biogeogr. Soc. Japan* **16–19**:271–277.

Iwai, T. (1967) "Structure and development of lateral line cupulae in teleost larvae." In *Lateral line detectors,* ed. Phyllis H. Cahn, pp. 27–44. Bloomington, Indiana University Press.

Jacobs, W. (1937) "Beobachtungen über das Schweben der Siphonophoren." *Z. vergl. Physiol.* **24** (4):583–601.

Jefferies, R. P. S. (1968) "The sybphylum Calcichordata (Jefferies, 1967) primitive fossil chordates with echinoderm affinities." *Bull. Br. Mus. nat. Hist. (Geol.)* **16**(6):241–339.

Jerlov, N. G. (1966) "Aspects of light measurements in the sea." In "Light as an ecological factor." *Symp. Br. ecol. Soc.* no. 6:91–98.

Jespersen, P. (1915) "Sternoptychidae (*Argyropelecus* and *Sternoptyx*)." *Rep. Danish oceanogr. Exped. Medit. 2 (Biol.) A.* **2**:1–41.

Jespersen, P., and A. V. Tåning (1926) "Mediterranean Sternoptychidae." *Rep. Danish oceanogr. Exped. Medit. 2 (Biol.) A.* **12**:1–59.

Johnson, F. H., H. Eyring, and M. J. Polissar (1954) *The kinetic basis of molecular biology.* New York, John Wiley.

Jones, F. R. H., and N. B. Marshall (1953) "The structure and functions of the teleostean swimbladder." *Biol. Rev.* **28**: 16–83.

Jones, G. M., and K. E. Spells (1963) "A theoretical and comparative study of the functional dependence of the semicircular canal upon its physical dimensions." *Proc. Roy. Soc. B.* **157** (968):403–419.

Jørgensen, C. Barker (1966) *Biology of suspension feeding.* Oxford, Pergamon Press.

References

Kawaguchi, K., and R. Marumo (1967) "Biology of *Gonostoma gracile* (Gonostomatidae) I. Morphology, life history and sex reversal." *Information Bulletin on planktology in Japan,* pp. 53–67. Commemoration number of Dr. Y. Matsue.

Knight-Jones, E. W., and E. Morgan (1966) "Responses of marine animals to changes in hydrostatic pressure." *Oceanogr. mar. Biol.* **4**:267–299.

Kuhn, W., A. Ramel, H. J. Kuhn, and E. Marti (1963) "The filling mechanism of the swimbladder." *Experientia* **19** (10):457–511.

Langer, Susanne K. (1967) *Mind: an essay on human feeling.* Vol. 1. Baltimore, The Johns Hopkins Press.

Lévi-Strauss, C. (1968) *Structural Anthropology.* London, The Penguin Press.

Lissmann, H. W. (1958) "On the function and evolution of electric organs in fish." *J. exp. Biol.* **35**:156–191.

——— (1961) "Ecological studies on gymnotids (in) Bioelectrogenesis." *Proceedings of the symposium on comparative bioelectrogenesis,* ed. Carlos Chagas and Antonio Paes de Carvalho, pp. 215–226. Elsevier.

Lissmann, H. W., and K. E. Machin (1958) "The mechanism of object location in *Gymnarchus niloticus* and similar fish." *J. exp. Biol.* **35**:451–486.

Lissmann, H. W., and H. O. Schwassmann (1965) "Activity rhythm of an electric fish, *Gymnorhamphichthys hypostomus* Ellis." *Z. vergl. Physiol.* **51**:153–171.

MacArthur, R. H., and J. H. Connell (1966) *The biology of populations.* New York, John Wiley.

McCutcheon, F. H. (1966) "Pressure sensitivity, reflexes and buoyancy responses in teleosts." *Anim. Behav.* **14** (2–3): 204–217.

McLaren, I. A. (1963) "Effects of temperature on growth of zooplankton and the adaptive value of vertical migration." *J. Fish. Res. Bd. Canada* **20**:685–727.

Madge, C. (1963) "Myth, metaphor and world picture." Manchester Literary and Philosophical Society, *Memoirs and Proceedings* **105**:1962–63.

Magnuson, J. J., and J. H. Prescott (1966) "Courtship, locomotion, feeding and miscellaneous behaviour of Pacific Bonito (*Sarda chiliensis*)." *Anim. Behav.* **14**:54–67.

Marshall, A. J. (1960) "Reproduction in male bony fish." In *Hormones in fish. Symp. Zool. Soc. Lond.* no. 1:137–151.

Marshall, N. B. (1954) *Aspects of deep sea biology.* Hutchinsons, London.

——— (1960) "Swimbladder structure of deep-sea fishes in relation to their systematics and biology." *Discovery Rep.* **31**: 1–122.

——— (1962) "The biology of sound-producing fishes." *Symp. Zool. Soc. Lond.* no. 7:45–60.

——— (1965a) "Systematic and biological studies of the macrourid fishes (Anacanthini-Teleostii)." *Deep-Sea Res.* **12**: 229–322.

——— (1965b) *The life of fishes.* Weidenfeld and Nicolson, London.

———(1966) "Fishes of the Western North Atlantic. Family Scopelosauridae." *Memoir Sears Foundation for Marine Research* **1** (5):194–204.

——— (1967) "The olfactory organs of bathypelagic fishes." *Symp. zool. Soc. Lond.* no. 19:57–70.

Marshall, N. B., and G. L. Thines (1958) "Studies of the brain, sense organs and light sensitivity of a blind cave fish (*Typhlogarra widdowsoni*) from Iraq." *Proc. zool. Soc. Lond.* **131** (3):441–456.

Marshall, N. B., and D. W. Bourne (1964) "A photographic survey of benthic fishes in the Red Sea and Gulf of Aden, with observations on their population density, diversity, and habits." *Bull. Mus. comp. Zool. Harv.* **132**(2):223–244.

Marshall, N. B., and A. V. Tåning (1966) "The bathypelagic fish, *Macrouroides inflaticeps* Smith and Radcliffe." *Dana Rep.* no. 69:1–6.

Mayr, Ernst (1963) *Animal species and evolution.* Cambridge, Mass., Harvard University Press.

——— (1965) *The Scientific Endeavour.* New York, The Rockefeller Institute Press.

Mead, G. W. (1960) "Hermaphroditism in archibenthic and pelagic fishes of the order Iniomi." *Deep-Sea Res.* **6**:234–235.

——— (1963) "Observations on fishes caught over the anoxic waters of the Cariaco Trench, Venezuela." *Deep-Sea Res.* **10** (3):251–257.

——— (1966) "Family Chlorophthalmidae." In "Fishes of the Western North Atlantic." *Mem. Sears Fdn. mar. Res.* no. 1m pt. 5:162–189.

Mead, G. W., E. Bertelsen, and D. M. Cohen (1964) "Reproduction among deep sea fishes." *Deep-Sea Res.* **II**: 569–596.

Meerloo, J. A. M. (1968) "The time sense in psychiatry." In *The voices of time,* ed. J. T. Fraser, pp. 235–252. London, The Penguin Press.

Menzies, R. J., and R. Y. George (1967) "A re-evaluation of the concept of hadal or ultra-abyssal fauna." *Deep-Sea Res.* **14** (6):703–723.

Miller, R. V. (1964) "The morphology and function of the pharyngeal organs in the clupeid, *Dorosoma petenense* (Günther)." *Chesapeake Sci.* **5**:194–199.

Morton, J. E. (1958) *Molluscs.* London, Hutchinson.

Morton, J., and M. Miller (1968) *The New Zealand sea shore.* London, Collins.

Moulton, J. M. (1960) "Swimming sounds and the schooling of fishes." *Biol. Bull. Woods Hole* **119**:210–223.

Muir, B. S., and J. I. Kendall (1968) "Structural modifications in the gills of tunas and other oceanic fishes." *Copeia* no. 2: 388–398.

Mukhacheva, V. A. (1966) "The composition of species of the genus *Cyclothone* (Pisces, Gonostomidae) in the Pacific Ocean." In *Fishes of the Pacific and Indian Oceans: Biology and distribution,* ed. T. S. Rass, pp. 98–146. Israel Program for Scientific Publications.

References

Munk, O. (1964) "Ocular degeneration in deep-sea fishes." *Galathea Rep.* **8**:21.

——— (1965) "*Omosudis lowei* Günther, 1887, a bathypelagic deep-sea fish with an almost pure-cone retina." *Vidensk. Medd. fra Dansk naturh. Foren* **128**:341–355.

——— (1966) "Ocular anatomy of some deep-sea teleosts." *Dana Rep.* no. 70:1–62.

Murray, P. D. F. (1936) *Bones*. Cambridge, England, Cambridge University Press.

Nafpaktitis, B. G. (1966) "Two new fishes of the myctophid genus *Diaphus* from the Atlantic Ocean." *Bull. Mus. comp. Zool. Harv.* **133**:401–424.

Napora, T. A. (1964) "The effect of hydrostatic pressure on the prawn, *Systellaspis debilis*." *Narragansett Mar. Lab. Occ. Publ.* **2**:92–94.

Nelson, G. J. (1967a) "Branchial muscles in some generalized teleostean fishes." *Acta Zoologica* **98**:277–288.

——— (1967b) "Epibranchial organs in lower teleostean fishes." *J. Zool. Lond.* **153**:71–89.

Nervi, P. L. (1965) "Is architecture moving towards unchangeable forms?" In *Structure in art and in science*, ed. Gyorgy Kepes, pp. 96–104. London, Studio Vista.

Neurath, H., K. A. Walsh, and W. P. Winter (1967) "Evolution of structure and function of proteases." *Science* **158**(3809): 1638–1644.

Nicol, J. A. C. (1958) "Observations on luminescence in pelagic animals." *J. mar. biol. Ass. U.K.* **37**:705–752.

——— (1967) "The luminescence of fishes." *Symp. zool. soc. Lond.* no. 19:27–55.

Nielsen, J. G. (1966) "Synopsis of the Ipnopidae (Pisces, Iniomi) with description of two new abyssal species." *Galathea Rep.* **8**:49–75.

Nielsen, J. G., and O. Munk (1964) "A hadal fish (*Bassogigas profundissimus*) with a functional swimbladder." *Nature* **204**(4958):594–595.

Nieuwenhuys, R. (1962) "Trends in the evolution of the actinopterygian forebrain." *J. Morph.* **3**(1):69–88.

Nursall, J. R. (1963a) "The hypurapophysis: An important element of the caudal skeleton." *Copeia*, pp. 458–459.

——— (1963b) "The caudal musculature of *Hoplopagrus guntheri* Gill (Perciformes: Lutianidae)." *Canad. J. Zool.* **41**:865–880.

O'Day, W. T., and B. G. Nafpaktitis (1967) "A study of the effects of expatriation on the gonads of two myctophid fishes in the North Atlantic Ocean." *Bull. Mus. Comp. Zool. Harv.* **136**:77–90.

Odum, E. P. (1959) *Fundamentals of ecology*. Philadelphia, W. B. Saunders Company.

Orton, Grace L. (1953) "The systematics of vertebrate larvae." *Systematic Zool.* **2**:63–75.

Orton, J. H. (1914) "On the ciliary mechanisms in brachiopods and some polychaetes, with a comparison of ciliary mechanisms on the gills of molluscs, Protochordata, brachiopods and cryptocephalous polychaets, and an

account of the endostyle of *Crepidula* and its allies."
J. mar. biol. Ass. U.K. 10 (2):283–311.

Pantin, C. F. A. (1951) "Organic design." *Advmt. Sci. Lond.* **8** (30:138–150.

———— (1965) "Life and the conditions of existence." In *Biology and personality,* ed. I. T. Ramsey, pp. 83–106. Oxford, Blackwell.

Patterson, C. (1964) "A review of Mesozoic acanthopterygian fishes with special reference to those of the English Chalk." *Phil. Trans. Roy. Soc.* ser. B. **247**:213–482.

Patterson, C. (1967) "Are the teleosts a polyphyletic group?" *Colloques int. Cent. natn. Rech. scient.* (Paris) **163**:93–109.

Pearcy, W. G., and R. M. Laurs (1966) "Vertical migration and distribution of mesopelagic fishes off Oregon." *Deep-sea Res.* **13**:153–165.

Pickwell, G. V. (1967) "Gas and bubble production by siphonophores." *Naval Undersea Warfare Center,* San Diego, Calif. TP.8.

Pfeiffer, W. (1962) "The fright reaction of fish." *Biol. Rev.* **37**: 495–511.

———— (1964) "Equilibrium orientation in fish." *Int. Rev. Gen. Expl. Zool.* **1**:77–111.

Popovici, Z. (1931) "Untersuchungen über die Seitenlinie der Knockenfische." *Jena Z. Naturw.* **65**:1–244.

Poulson, T. L. (1963) "Cave adaptation in Amblyopsid fishes." *Amer. midl. Nat.* **70**:257–290.

Prosser, C. Ladd (1965) "Levels of biological organization and their physiological significance." In *Ideas in modern biology,* ed. John A. Moore, pp. 357–390. Proc. Vol. 6 XVI International Congress of Zoology. New York, The Natural History Press.

Rass, T. S. (1966) "Changes in eye size and body coloration in secondary deep-sea fishes." In *Fishes of the Pacific and Indian Oceans, Biology and distribution.* Ed. T. S. Rass, pp. 1–9. Israel Program for scientific publications.

Rayner, D. H. (1948) "The structure of certain Jurassic holosteans with special reference to their neurocrania." *Phil. Trans. R. Soc. B.* **233**:287–345.

Regan, C. T. (1907) "On the anatomy, classification and systematic position of the teleostean fishes of the suborder Allotriognathi." *Proc. zool. Soc. Lond.,* pp. 634–643.

———— (1929) Article on Fishes. *Encyclopaedia Britannica,* 14th ed.

———— (1936) "Pisces." In *Natural History,* ed. C. T. Regan, pp. 201–296. London, Ward, Lock & Co.

Reid, J. L. (1965) "Intermediate waters of the Pacific Ocean." *The Johns Hopkins Oceanographic Studies No. 2.* Baltimore, The Johns Hopkins Press.

Rofen, R. R. (1966) "Fishes of the Western North Atlantic. Family Paralepididae." *Memoir Sear Foundation for Marine Research* **1** (5):205–461.

Rosen, D. E. (1964) "The relationships and taxonomic position of the halfbeaks, killifishes, silversides and their relatives." *Bull. Amer. Mus. nat. Hist.* **127**:217–267.

References

Rosen D. E., and C. Patterson (1969) "The structure and relationships of the paracanthopterygian fishes." *Bull. Amer. Mus. nat. Hist.* **141**, art. 3:357–474.

Rosen, R. (1967) *Optimality principles in biology.* London, Butterworth.

Russell, E. S. (1946) *The directiveness of organic activities.* Cambridge, England, Cambridge University Press.

Schaeffer, B., and D. E. Rosen (1961) "Major adaptive levels in the evolution of the actinopterygian feeding mechanism." *Am. Zool.* **1**:187–204.

Scharrer, E. (1959) "General and phylogenetic interpretation of neuroendocrine interrelations." In *Comparative endocrinology,* ed. Aubrey Gorbman, pp. 233–249. New York, John Wiley.

Schlieper, M. C. (1968) "High pressure effects on marine invertebrates and fishes." *Mar. Biol.* (Berlin) **2**:5–12.

Schmidt-Nielsen, Bodil (1965) "Comparative morphology and the physiology of excretion." In *Ideas in modern biology,* ed. John A. Moore, pp. 391–425. Proc. Vol. 6. XVI International Congress of Zoology. New York, The Natural History Press.

Schneider, H. (1962) "The labyrinth of two species of drumfish (Sciaenidae)." *Copeia,* pp. 336–338.

Scholander, P. F. (1958) "Counter current exchange: A principle in biology." *V Hval. Skr. Nr.* **44**:1–24.

Schwartz, E. (1965) "Bau und Funktion der Seitenlinie des Streifenhechtlings (*Aplocheilus lineatus* Cuv u Val)." *Z. vergl. Physiol.* **50**:55–87.

——— (1967) "Analysis of surface-wave perception in some teleosts." In *Lateral line detectors,* ed. Phyllis H. Cahn, pp. 123–134. Bloomington, Indiana University Press.

Simpson, G. G. (1949) *The meaning of evolution.* New Haven, Yale University Press.

——— (1953) *The major features of evolution.* New York. Colombia University Press.

——— (1961) *Principles of animal taxonomy.* New York, Columbia University Press.

——— (1964) "Organisms and molecules in evolution." *Science* **146** (3651).

Smith, C. S. (1965) "Structure, substructure and superstructure." In *Structure in art and in science,* ed. Gyorgy Kepes, pp. 29–41. London, Studio Vista.

Springer, S. (1960) "Natural history of the sandbar shark *Eulamia milberti.*" *U.S. Fish and Wildlife Service, Fish. Bull,* **61** No. 178:1–38.

Stefanelli, A. (1962) "Neurología ecológica de los centros estáticos de los vertebratos." *Biológica* (Santiago). No. 33.

Stewart, K. W. (1962) "Observations on the morphology and optical properties of the adipose eyelid of fishes." *J. Fish. Res. Bd. Can.* **19**:1161–1162.

Swedmark, B. (1964) "The interstitial fauna of marine sand." *Biol. Rev.* **39** (1):1–42.

Swallow, Mary (1962) "Deep currents in the oceans." *Discovery.* **23** (6):17–22.

Szabo, T. (1965) "Sense organs of the lateral line system in some electric fish of the Gymnotidae, Mormyridae and Gymnarchidae." *J. Morph.* **117**:229–250.

Tchindonova, J. G. (1966) "On concentrations of bathypelagic fishes and other animals of the Atlantic Ocean recorded by echo-sounding." *Second International Oceanographic Congress.* Moscow, Nauka. (Abstract.)

Teal, J. M., and F. G. Carey (1967) "Effects of pressure and temperature on the respiration of euphausiids." *Deep-Sea Res.* **14** (6):725–733.

Te Winkel, Lois E. (1935) "A study of *Mystichthys luzonensis* with special reference to conditions correlated with reduced size." *J. Morph.* **58** (2):463–536.

Thinés, G. (1955) "Les poissons aveugles: (1) Origines, taxonomie, répartition géographique, comportement." *Ann. Soc. roy. Zool. Belg.* **86**:1–128.

Thompson, D'Arcy, W. (1961) *On growth and form,* ed. John Tyler Bonner. Cambridge, England, Cambridge University Press.

Thorson, G. (1946) "Reproduction and larval development of Danish marine bottom invertebrates, with special reference to the planktonic larvae in the sound (Öresund)." *Medd. Komm. Danmarks Fisk Havunders. Ser. Plankton.* **4** no. 1:1–523.

Tiegs, O. W., and S. M. Manton (1958) "The evolution of the Arthropoda." *Biol. Rev.* **33**:255–337.

Tuge, H., K. Uchihashi, and H. Shimamura (1968) *An atlas of the brains of fishes of Japan.* Tokyo, Tsukiji Shokan Publishing Co.

Tyler, J. C. (1962) "*Triodon bursarius:* A plectognath fish connecting the Sclerodermi and Gymnodontes." *Copeia* No. 4:793–801.

Vilter, V. (1954) "Différenciation fovéale dans l'appareil visuel d'un poisson abyssal, le *Bathylagus benedicti.*" *C. R. Soc. Biol.* (Paris). **148**:59.

Vinogradov, M. E. (1968) Vertical distribution of the oceanic zooplankton. Moscow, Publishing House "Nauka."

Vinogradova, N. G. (1962) "Some problems of the study of deep-sea bottom fauna." *J. oceanogr. Soc. Japan.* 20th anniversary volume. pp. 724–741.

Voss, G. L. (1967) "The pelagic mid-water fauna of the eastern tropical Atlantic with special reference to the Gulf of Guinea." Reprint of paper given at UNESCO symposium on the marine resources of the Tropical Atlantic (1966).

Wald, G. (1965) "The origins of life." In *The scientific Endeavour,* pp. 113–134. New York, The Rockefeller Institute Press.

Walls, G. L. (1942) "The vertebrate eye and its adaptive radiation." *Cranbrook Institute of Science, Bull.* no. 19:XIV–785.

Walters, V., and H. L. Fierstein (1964) "Measurements of swimming speeds of yellowfin tuna and wahoo." *Nature* **202**:208–9.

Wardlaw, C. W. (1965) *Organization and evolution in plants.* London, Longmans.

References

Waterman, T. H. (1948) "Studies on the deep-sea angler-fishes (Ceratioidea) III. The comparative anatomy of *Gigantactis longicirra* Waterman." *J. Morph.* **82**:81–149.

Watts, E. H. (1961) "The relationship of fish locomotion to the design of ships." *Symp. zool. Soc. Lond.* No. 5:37–42.

Weitzman, S. H. (1960) "The phylogenetic relationships of *Triportheus*, a genus of South American characid fishes." *Stanford ichthyol. Bull.* **7**:239–244.

—— (1962) "The osteology of *Brycon meeki*, a generalized characid fish, with an osteological definition of the family." *Stanford ichthyol. Bull.* **8**:1–77.

—— (1967) "The origin of the stomiatoid fishes with comments on the classification of Salmoniform fishes." *Copeia.* No. 3:507–540.

Wells, M. J. (1962) *Brain and behaviour in cephalopods.* London, Heinemann.

Whyte, L. L. (1965a) *Internal factors in evolution.* London. Tavistock Publications.

—— (1965b) "Atomism, Structure and Form." In *Structure in art and in science,* ed. Gyorgy Kepes, pp. 20–28. London, Studio Vista.

Willemse, J. J. (1966) "Functional anatomy of the myosepta in fishes." *K. Nederl. Wetensch. Prov.* **69** (1):58–63.

Wittenberg, J. B. and B. A. Wittenberg (1962) "Active secretion of oxygen into: The eye of fish." *Nature, Lond.* **194**:106–107.

Wolff, T. (1961) "The deepest recorded fishes." *Nature* **190** (4772):283.

Wynne-Edwards, V. C. (1962) *Animal dispersion in relation to social behaviour.* London, Oliver and Boyd.

Yasargil, G. M., and J. Diamond (1968) "Startle response in teleost fish: An elementary circuit for neural discrimination." *Nature* **220** (5164):241–273.

Young, J. Z. (1964) *A model of the brain.* Oxford, Clarendon Press.

Zenkevitch, L. A. (1945) "The evolution of animal locomotion." *J. Morph.* **77**:1–52.

—— (1963) *Biology of the seas of the U.S.S.R.* London, Allen and Unwin.

Zobell, C. E., and R. Y. Morita (1956) "Bacteria in the deep sea." *The Galathea Deep Sea Expedition 1950-52,* ed. A. F. Bruun et al., pp. 201–210. London, Allen and Unwin.

Zuckerkandl, E., and L. Pauling (1965) "Evolutionary divergence and convergence." In *Evolving genes and proteins,* ed. V. Bryson and H. J. Vogel, pp. 97–166. New York, Academic Press.

Index

Index

Index

Index

Index

Index

Index

Index

Index

skeleton, 9; living spaces, 1; movements of fins, 10; number of caudal rays, 32; number of individuals, 1; relatively small swimbladder and buoyancy adjustment, 8–9; schooling species, 14; sense organs, 12–15; soft-finned species dominant in deep ocean, 18; sound signals, 14; spiny-finned species dominant in shallow seas, 17–18; structure and function of cerebellum, 14; strutting of bones, 9; swimbladder and buoyancy, 7–9; visual signals, 14

Temperature profile, of deep ocean, 37

Tetraodon fluviatilis (puffer-fish), fovea in retina, 115

Te Winkel, Lois. E., 4

Thermocline: permanent, 37; "seasonal," 37

Thermosphere, 37

Thinés, G., 48

Tholichthys stage of chaetodont fishes, 101

Thompson, D'Arcy, W., 165

Thoracocharax, flying characid fishes, 16

Thorson, G., 102

Thrissobrycon (characid fish), convergence with herring-like fishes, 16

Thunnus (scombroid fish): convergences with isurid sharks, 138; form of gill filaments, 138; subcutaneous vascular plexus, 140; *T. albacares*, 25, 26

Thysanopoda spp. (euphausiid shrimps), organization, 161

Tiegs, O. W., and S. M. Manton, 102

Toadfish (*Opsanus tau*), relative gill surface, 148

Top-minnows (Poeciliidae), adaptations for surface feeding

Torpedo-rays (Torpedinidae), electric organs, 109

Trachinus (weever-fish) fovea in retina, 115; *T. radiatus*, reduced Mauthnerian system, 29

Trachypterus (trachypterid fish), epipelagic, 137

Trachyrhynchus (macrourid fish), organization, 75

Transparency of ocean, 38

Trichiuroid fishes, fixed jaws, 10

Trigger-fishes, fovea in retina, 115

Tripod fishes (Bathypteroidae), deep-sea benthic fishes, 48

Triportheus (characid fish), keeled thorax and winged pectorals, 136

Tuge, H., K. Uchihashi, and H. Shimamura, 120

Tunas: body form and musculature, 24–25; convergences with isurid sharks, 138–140

Turbot (*Scophthalmus maximus*), caudal structure and movement in young, 31

Twilight zone, 35

Tyler, J. C., 32

Typhlichthys (amblyopsid fish), free-ending neuromasts, 54

Typhlonus (brotulid fish), regressed eyes, 48

Valenciennellus (stomiatoid fish), silvery sides, 62

Vampyroteuthis infernalis (cephalopod), in oxygen-minimum layer, 3

Vertical migrations: advantages, 66; mesopelagic fishes, 9, 64–66, 142; metabolic advantages, 76–77

Vilter, V., 41

Vinciguerria (stomiatoid fish), 37; organization, 68; red muscle, 78; silvery sides, 62; ventral photophores, 63; vertical migrations in Red Sea, 66; *V. attenuata*, incipient tubular eyes, 44

Vinogradova, N. G., 80

Visual cell mosaics, in teleosts, 20

Von Arx, W. S., 37

Voss, G. L., 96

Wahoo (*Acanthocybium solandri*), speed, 26

Wald, G., 107

Walls, G. L., 41, 49, 112, 113, 115, 148, 157

Warm-blooded fishes, 139

Waterman, T. H., 70

Watts, E. H., 27, 138

Wavyback skipjack (*Euthynnus affinis*), analysis of locomotion, 25

Weever (*Trachinus radiatus*): fovea in retina, 115; Mauthnerian system, 29

Weitzman, S. H., 16, 17, 135, 136